The Search for Certainty

A Journey Through the History of Mathematics, 1800–2000

Edited by Frank J. Swetz

Dover Publications, Inc.
Mineola, New York

Bibliographical Note

This Dover edition, first published in 2012, contains a new selection from
the work *From Five Fingers to Infinity: A Journey Through the History of
Mathematics,* originally published by Open Court Publishing Company,
Chicago, in 1994.

Historical Exhibit 5 and Chapters 3 and 6 have been added to this
Dover edition.

International Standard Book Number

ISBN-13: 978-0-486-47442-7
ISBN-10: 0-486-47442-9

Manufactured in the United States by Courier Corporation
47442901
www.doverpublications.com

Table of Contents

Preface

*W*elcome! You are about to embark on a journey—an intellectual excursion through the history of mathematics. Although this journey is limited to the time between 1800 to 2000, it will be eventful and informative. In mathematical history this period holds special significance: it marks the rise of modern mathematics. As with any journey, an itinerary is determined by many factors. In a physical journey this may include travel time available, distance, and cost. In an intellectual journey the itinerary is determined by the series of particular events or scenic attractions that mark important conceptual advances; issues that reflect the mathematical climate of the time and, of course, the availability of appropriate articles that discuss these topics. As your tour guide, I have tried to select events that impact the understanding of how mathematics is developed. Historical exhibits "snapshot" some prominent features that warrant consideration on their own. As with any journey experiencing events is transitory, facts are encountered, and impressions conveyed—the distance between encounters might vary according to the terrain. In your travels, one day you may move slowly taking in more scenery; on another, you hurry through only seeing the highlights. It is similar traveling through the history of mathematics, some events, due to their importance and/or impact are considered in detail, and others of less importance or overwhelming complexity are merely surveyed.

The period for our travels is limited to two centuries whereas records of human activity with mathematics extend back several millennia. It is hoped that in the following months other travel itineraries, volumes in this series, will complete this journey of understanding by exploring earlier centuries.

FRANK J. SWETZ
Harrisburg, Pennsylvania
December 2008

*F*OLLOWING THE ADVENT OF CALCULUS in the seventeenth century came a period of experimentation and mathematical exploitation. For the next hundred years, mathematicians revelled in the use of the powerful new analytic methods supplied to them by Newton and Leibniz. Under a prevailing spirit of mathematical adventurism, many results were obtained the implications of which were not fully understood at the time. As the mathematical climate of Europe was rapidly changing, the social and intellectual climate in general was also in a state of flux. The French Revolution (1789) and the following Napoleonic period (1804–1815) encouraged a new critical outlook on the sciences, education, and the relation of these institutions to society. Science was becoming democratized. Europe, as a whole, was beginning to experience the impact of industrialization and the rise of technological innovations. Mathematical answers were sought to an increasing number of technological and sociological questions. The realm of mathematics moved from the comfortable isolation of royal courts and elite academies to the lecture halls of universities and research institutes, which were more demanding testing grounds for mathematical ideas. Such ideas now had to be effectively communicated to large receptive audiences and attention was paid to producing adequate textbooks for this task. No more could a single individual possess a universal knowledge of mathematics—the subject had become too broad and complex. The different areas of mathematics came to be classified as either "pure" or "applied" and the constrained specialist appeared on the scene.

The development of mathematics in the nineteenth century underwent alternate phases of consolidation, rigorization, and fragmentation. Outstanding questions from Greek antiquity were resurrected, examined, and, in some cases, put to rest. By the 1880s, the three classical Greek problems—duplication of the cube, angle trisection, and the quadrature of the circle—were proven to be unsolvable under the restriction of ruler and compass construction. Constructibility of regular polygons was also explored using modern techniques and the class of constructible polygons extended. The nagging question of the independence of Euclid's fifth postulate was contested and the postulate was found to be independent within the Euclidean system. Thus, alternate theories of "parallelism" were permitted giving rise to non-Euclidean geometries. Strange new mathematical worlds were appearing.

A reorganization of the calculus by such mathematicians as Augustin Cauchy (1789–1848), Karl Weierstrass (1815–1897), and Bernard Bolzano (1781–1848) amended the haphazard advances of the eighteenth century and set calculus on a sound logical foundation. These efforts to strengthen mathematics often led to more perplexing and intriguing questions and problems, such as: "What is a function?" "How could the concept of continuity be formalized?" "What was the structure of the real numbers?" Number systems were given special attention. Both geometric and analytic interpretations were provided for the complex numbers. Building upon complex number theory, the Irish mathematician William Hamilton (1805–1865) sought to develop meaningful mathematical operations in a four-dimensional space. His efforts resulted in the discovery of quaternions and noncommutative mathematical operations. Algebra was no longer merely an "arithmetic of unknowns" but became a study of the structure of mathematics itself. The German mathematician Georg Cantor (1845–1918) grappled with the concept of infinity (a concept that had defeated Zeno and his fellow sophists) and produced workable theories of classification and manipulation for transfinite numbers. Mathematics was changing so fast that mathematicians were compelled to ask the ultimate question—"What is mathematics?"

Towards the turn of the century, three schools of mathematical thought arose to confront this question. The first of these, formalism, insisted that all mathematics could be treated symbolically and depended on the use of a rigorous axiomatic approach to formulate concepts; new theories then emerged from a sequence of deductive steps from the original axioms. The validity of formalism rested on the consistency of the basic axiomatic system—if the system was consistent and used correctly then correct mathematics would result. Formalism depended on logic but not as the primary means of understanding mathematics. In contrast, a second philosophical school of thought, logicism, contended that mathematics was simply logic. The chief proponents of this theory were the British philosophers Bertrand Russell and Alfred North Whitehead, who published their views in the monumental work *Principia Mathematica* (1910–1913). Finally, the third school of thought was championed by the Dutch mathematician L. E. J. Brouwer, who felt that mathematics was independent of both language and logic. He considered mathematical concepts mental constructs formed by a finite sequence of intuitive steps that could be traced back to the natural numbers. The idea that mathematics is basically intuitive is called intuitionism. Each of these schools eventually encountered difficulties with its doctrines. For example, in 1931 a brilliant young German mathematician, Kurt Gödel, demonstrated that consistency could never be adequately proven within a given formal system, thus casting doubt on the very basis of formalism. To this day, no school of thought has definitively answered the question, "What is mathematics?"

Although a state of uncertainty exists to plague mathematicians, both the nature and extent of that uncertainty is being reduced. In particular, the advent and continuing development of electronic computers and electronic computing techniques are rapidly pushing the frontiers of mathematics into new realms. In 1976, Kenneth Appel and Wolfgang Haken of the University of Illinois used 1000 hours of computer time to devise 1,936 configurations to prove the Four-Color-Conjecture. This work marks the first instance of a computer derived proof in mathematics. The largest known prime number of 65,087 digits has been discovered by computer (1989) as well as a billion decimal place estimate for the value of π. While the usefulness of such knowledge is debatable, its accomplishment testifies to the ongoing challenges of mathematics and how they are being met.

1

The Changing Concept of Change: The Derivative from Fermat to Weierstrass

JUDITH V. GRABINER

SOME YEARS AGO while teaching the history of mathematics, I asked my students to read a discussion of maxima and minima by the seventeenth-century mathematician, Pierre Fermat. To start the discussion, I asked them, "Would you please define a relative maximum?" They told me it was a place where the derivative was zero. "If that's so," I asked, "then what is the definition of a relative minimum?" They told me, *that's* a place where the derivative is zero. "Well, in that case," I asked, "what is the difference between a maximum and a minimum?" They replied that in the case of a maximum, the second derivative is negative.

What can we learn from this apparent victory of calculus over common sense?

I used to think that this story showed that these students did not understand the calculus, but I have come to think the opposite: they understood it very well. The students' answers are a tribute to the power of the calculus in general, and the power of the concept of derivative in particular. Once one has been initated into the calculus, it is hard to remember what it was like *not* to know what a derivative is and how to use it, and to realize that people like Fermat once had to cope

with finding maxima and minima without knowing about derivatives at all.

Historically speaking, there were four steps in the development of today's concept of the derivative, which I list here in chronological order. The derivative was first *used*; it was then *discovered*; it was then *explored and developed*; and it was finally *defined*. That is, examples of what we now recognize as derivatives first were used on an ad hoc basis in solving particular problems; then the general concept lying behind these uses was identified (as part of the invention of the calculus); then many properties of the derivative were explained and developed in applications both to mathematics and to physics; and finally, a rigorous definition was given and the concept of derivative was embedded in a rigorous theory. I will describe the stops, and give one detailed mathematical example from each. We will then reflect on what it all means—for the teacher, for the historian, and for the mathematician.

The Seventeenth-Century Background

Our story begins shortly after European mathematicians had become familiar once more with Greek mathematics, learned Islamic algebra, synthesized the two traditions, and struck out on their

Reprinted from *Mathematics Magazine* 56 (Sept., 1983): 195–206; with permission of the Mathematical Association of America.

own. François Vieta invented symbolic algebra in 1591; Descartes and Fermat independently invented analytic geometry in the 1630s. Analytic geometry meant, first, that curves could be represented by equations; conversely, it meant also that every equation determined a curve. The Greeks and Muslims had studied curves, but not that many—principally the circle and the conic sections plus a few more defined as loci. Many problems had been solved for these, including finding their tangents and areas. But since any equation could now produce a new curve, students of the geometry of curves in the early seventeenth century were suddenly confronted with an explosion of curves to consider. With these new curves, the old Greek methods of synthetic geometry were no longer sufficient. The Greeks, of course, had known how to find the tangents to circles, conic sections, and some more sophisticated curves such as the spiral of Archimedes, using the methods of synthetic geometry. But how could one describe the properties of the tangent at an arbitrary point on a curve defined by a ninety-sixth degree polynomial? The Greeks had defined a tangent as a line which touches a curve without cutting it, and usually expected it to have only one point in common with the curve. How then was the tangent to be defined at the point $(0, 0)$ for a curve like $y = x^3$ (Figure 1), or to a point on a curve with many turning points (Figure 2)?

The same new curves presented new problems to the student of areas and arc lengths. The Greeks had also studied a few cases of what they called "isoperimetric" problems. For example, they asked: of all plane figures with the same perimeter, which one has the greatest area? The circle, of course, but the Greeks had no general method for solving all such problems. Seventeenth-century mathematicians hoped that the new symbolic algebra might somehow help solve all problems of maxima and minima.

Thus, though a major part of the agenda for seventeenth-century mathematicians—tangents, areas, extrema—came from the Greeks, the subject matter had been vastly extended, and the solutions would come from using the new tools: symbolic algebra and analytic geometry.

Finding Maxima, Minima, and Tangents

We turn to the first of our four steps in the history of the derivative: its *use*, and also illustrate some of the general statements we have made. We shall look at Pierre Fermat's method of finding maxima and minima, which dates from the 1630s [8]. Fermat illustrated his method first in solving a simple problem, whose solution was well known: *Given a line, to divide it into two parts so that the product of the parts will be a maximum.* Let the

FIGURE I

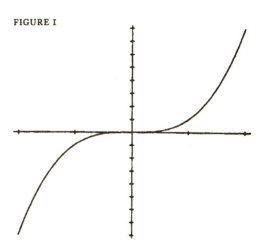

FIGURE 2

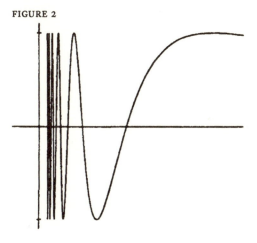

FIGURE 3

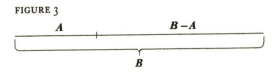

length of the line be designated B and the first part A (Figure 3). Then the second part is $B - A$ and the product of the two parts is

$$A(B - A) = AB - A^2. \qquad (1)$$

Fermat had read in the writings of the Greek mathematician Pappus of Alexandria that a problem which has, in general, two solutions will have only one solution in the case of a maximum. This remark led him to his method of finding maxima and minima. Suppose in the problem just stated there is a second solution. For this solution, let the first part of the line be designated as $A + E$; the second part is then $B - (A + E) = B - A - E$. Multiplying the two parts together, we obtain for the product

$$BA + BE - A^2 - AE - EA - E^2 =$$
$$AB - A^2 - 2AE + BE - E^2. \qquad (2)$$

Following Pappus's principle for the maximum, instead of two solutions, there is only one. So we set the two products (1) and (2) "sort of" equal; that is, we formulate what Fermat called the pseudo-equality:

$$AB - A^2 = AB - A^2 - 2AE + BE - E^2.$$

Simplifying, we obtain

$$2AE + E2 = BE$$

and

$$2A + E = B.$$

Now Fermat said, with no justification and no ceremony, "suppress E." Thus he obtained

$$A = B/2,$$

which indeed gives the maximum sought. He concluded, "We can hardly expect a more general method." And, of course, he was right.

Notice that Fermat did not call E infinitely small, or vanishing, or a limit; he did not explain why he could first divide by E (treating it as non-zero) and then throw it out (treating it as zero). Furthermore, he did not explain what he was doing as a special case of a more general concept, be it derivative, rate of change, or even slope of tangent. He did not even understand the relationship between his maximum-minimum method and the way one found tangents; in fact he followed his treatment of maxima and minima by saying that the same method—that is, adding E, doing the algebra, then suppressing E—could be used to find tangents [8, p. 223].

Though the considerations that led Fermat to his method may seem surprising to us, he did devise a method of finding extrema that worked, and it gave results that were far from trivial. For instance, Fermat applied his method to optics. Assuming that a ray of light which goes from one medium to another always takes the quickest path (what we now call the Fermat least-time principle), he used his method to compute the path taking minimal time. Thus he showed that his least-time principle yields Snell's law of refraction [7] [12, pp. 387–390].

Though Fermat did not publish his method of maxima and minima, it became well known through correspondence and was widely used. After mathematicians had become familiar with a variety of examples, a pattern emerged from the solutions by Fermat's method to maximum-minimum problems. In 1659, Johann Hudde gave a general verbal formulation of this pattern [3, p. 186], which, in modern notation, states that, given *a polynomial of the form*

$$y = \sum_{k=0}^{n} a_k x^k,$$

there is a maximum or minimum when

$$\sum_{k=1}^{n} k a_k x^{k-1} = 0.$$

Of even greater interest than the problem of extrema in the seventeenth century was the

FIGURE 4

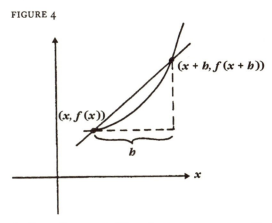

finding of tangents. Here the tangent was usually thought of as a secant for which the two points came closer and closer together until they coincided. Precisely what it meant for a secant to "become" a tangent was never completely explained. Nevertheless, methods based on this approach worked. Given the equation of a curve

$$y = f(x),$$

Fermat, Descartes, John Wallis, Isaac Barrow, and many other seventeenth-century mathematicians were able to find the tangent. The method involves considering, and computing, the slope of the secant,

$$\frac{f(x+h) - f(x)}{h},$$

doing the algebra required by the formula for $f(x + h)$ in the numerator, then dividing by h. The diagram in Figure 4 then suggests that when the quantity h vanishes, the secant becomes the tangent, so that neglecting h in the expression for slope of the secant gives the slope of the tangent. Again, a general pattern for the equations of slopes of tangents soon became apparent, and a rule analogous to Hudde's rule for maxima and minima was stated by several people, including René Sluse, Hudde, and Christiaan Huygens [3, pp. 185–186].

By the year 1660, both the computational and the geometric relationships between the problem of extrema and the problem of tangents were clearly understood; that is, a maximum was found by computing the slope of the tangent, according to the rule, and asking when it was zero. While in 1660 there was not yet a general concept of derivative, there was a general method for solving one type of geometric problem. However, the relationship of the tangent to other geometric concepts—area, for instance—was not understood, and there was no completely satisfactory definition of tangent. Nevertheless, there was a wealth of methods for solving problems that we now solve by using the calculus, and in retrospect, it would seem to be possible to gereralize those methods. Thus in this context it is natural to ask, how did the derivative as we know it come to be?

It is sometimes said that the idea of the derivative was motivated chiefly by physics. Newton, after all, invented both the calculus and a great deal of the physics of motion. Indeed, already in the Middle Ages, physicists, following Aristotle who had made "change" the central concept in his physics, logically analyzed and classified the different ways a variable could change. In particular, something could change uniformly or nonuniformly; if nonuniformly, it could change uniformly-nonuniformly or nonuniformly-nonuniformly, etc. [3, pp. 73–74]. These medieval classifications of variation helped to lead Galileo in 1638, without benefit of calculus, to his successful treatment of uniformly accelerated motion. Motion, then, could be studied scientifically. Were such studies the origin and purpose of the calculus? The answer is no. However plausible this suggestion may sound, and however important physics was in the later development of the calculus, physical questions were in fact neither the immediate motivation nor the first application of the calculus. Certainly they prepared people's thoughts for some of the properties of the derivative, and for the introduction into mathematics of the concept of change. But immediate motivation for the general concept of derivative—as opposed to specific examples like speed or slope of tangent—did not come from physics. The first problems to be solved, as well as the first applications, occurred in mathematics, especially geometry (see [1, chapter 7]; see also [3; chapters 4–5], and, for Newton, [17]). The concept of derivative then developed gradually, together with the ideas of extrema, tangent, area, limit, continuity, and function, and it interacted with these ideas in some unexpected ways.

Tangents, Areas, and Rates of Change

In the latter third of the seventeenth century, Newton and Leibniz, each independently, invented the calculus. By "inventing the calculus" I mean that they did three things. First, they took the wealth of methods that already existed for finding tangents, extrema, and areas, and they subsumed all these methods under the heading of two general concepts, the concepts which we now call *derivative* and *integral*. Second, Newton and Leibniz each worked out a notation which made it easy, almost automatic, to use these general concepts. (We still use Newton's $\dot{x}$ and we still use Leibniz's dy/dx and $\int y\,dx$.) Third, Newton and Leibniz each gave an argument to prove what we now call the Fundamental Theorem of Calculus: the derivative and the integral are mutually inverse. Newton called our "derivative" a *fluxion*—a rate of flux or change; Leibniz saw the derivative as a ratio of infinitesimal differences and called it the *differential quotient*. But whatever terms were used, the concept of derivative was now embedded in a general subject—the calculus—and its relationship to the other basic concept, which Leibniz called the integral, was now understood. Thus we have reached the stage I have called *discovery*.

Let us look at an early Newtonian version of the Fundamental Theorem [13, sections 54–5, p. 23]. This will illustrate how Newton presented the calculus in 1669, and also illustrate both the strengths and weaknesses of the understanding of the derivative in this period.

Consider with Newton a curve under which the area up to the point $D = (x, y)$ is given by z (see Figure 5). His argument is general: "Assume any

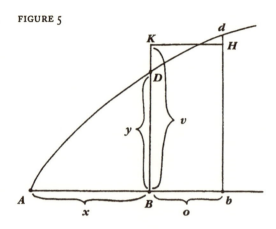

FIGURE 5

relation betwixt x and z that you please"; he then proceeded to find y. The example he used is

$$z = \frac{n}{m+n}\, ax^{(m+n)/n};$$

however, it will be sufficient to use $z = x^3$ to illustrate his argument.

In the diagram in Figure 5, the auxiliary line bd is chosen so that $Bb = o$, where o is not zero. Newton then specified that $BK = v$ should be chosen so that area $BbHK$ = area $BbdD$. Thus ov = area $BbdD$. Now, as x increases to $x + o$, the change in the area z is given by

$$z(x+o) - z(x) = x^3 + 3x^2 o + 3xo^2 + o^3 - x^3 = 3xo^2o + 3xo^2 + o^3,$$

which, by the definition of v, is equal to ov. Now since $3x^2 o + 3xo^2 + o^3 = ov$, dividing by o produces $3x^2 + 3ox + o^2 = v$. Now, said Newton, "If we suppose Bb to be diminished infinitely and to vanish, or o to be nothing, v and y in that case will be equal and the terms which are multiplied by o will vanish: so that there will remain . . . "

$$3x^2 = y.$$

What has he shown? Since $(z(x+o) - z(x))/o$ is the rate at which the area z changes, that rate is given by the ordinate y. Moreover, we recognize that $3x^2$ would be the slope of the tangent to the curve $z = x^3$. Newton went on to say that the argument can be reversed; thus the converse holds too. We see that derivatives are fundamentally involved in areas as well as tangents, so the concept of derivative helps us to see that these two problems are mutually inverse. Lebniz gave analogous arguments on this same point (see, e.g. [16, pp. 282–284]).

Newton and Leibniz did not, of course, have the last word on the concept of derivative. Though each man had the most useful properties of the concept, there were still many unanswered questions. In particular, what, exactly, is a differential quotient? Some disciples of Leibniz, notably Johann Bernoulli and his pupil the Marquis de l'Hospital, said a differential quotient was a ratio of infinitesimals; and after all, that is the way it was calculated. But infinitesimals, as seventeenth-century mathematicians were well aware, do not obey the Archimedean axiom. Since the Archimedean axiom was the basis for the Greek theory of ratios, which was, in turn, the basis of arithmetic, algebra, and geometry for seventeenth-century mathematicians, non-Archimedean objects were viewed with some suspicion. Again, what is a fluxion? Though it can be understood intuitively as a velocity, the proofs Newton gave in his 1671 *Method of Fluxions* all involved an "indefinitely small quanity o," [14, pp. 32–33] which raises many of the same problems that the o which "vanishes" raised in the Newtonian example of 1669 we saw above. In particular, what is the status of that little o? Is it zero? If so, how can we divide by it? If it is not zero, aren't we making an error when we throw it away? These questions had already been posed in Newton's and Leibniz's time. To avoid such problems, Newton said in 1687 that quantities defined in the way that $3x^2$ was defined in our example were the *limit* of the ratio of vanishing increments. This sounds good, but Newton's understanding of the term "limit" was not ours. Newton in his *Principia* (1687) described limits as "ultimate ratios"—that is, the value of the ratio of those vanishing quantities just when they are vanishing. He said, "Those ultimate ratios with which quantities vanish are not truly

the ratios of ultimate quantities, but limits towards which the ratios of quantities decreasing without limit do always converge; and to which they approach nearer than by any given difference, but never go beyond, nor in effect attain to, till the quantities are diminished in infinitum" [15, Book I, Scholium to Lemma XI, p. 39].

Notice the phrase "but never go beyond"—so a variable cannot oscillate about its limit. By "limit" Newton seems to have had in mind "bound," and mathematicians of his time often cite the particular example of the circle as the limit of inscribed polygons. Also, Newton said, "nor . . . attain to, till the quantities are diminished in infinitum." This raises a central issue: it was often asked whether a variable quantity ever actually reached its limit. If it did not, wasn't there an error? Newton did not help clarify this when he stated as a theorem that "Quantities and the ratios of quantities which in any finite time converge continually to equality, and before the end of that time approach nearer to each other than by any given difference, become ultimately equal" [15, Book I, Lemma I, p. 29]. What does "become ultimately equal" mean? It was not really clear in the eighteenth century, let alone the seventeenth.

In 1734, George Berkeley, Bishop of Cloyne, attacked the calculus on precisely this point. Scientists, he said, attack religion for being unreasonable; well, let them improve their own reasoning first. A quantity is either zero or not; there is nothing in between. And Berkeley characterized the mathematicians of his time as men "rather accustomed to compute, than to think" [2].

Perhaps Berkeley was right, but most mathematicians were not greatly concerned. The concepts of differential quotient and integral, concepts made more effective by Leibniz's notation and by the Fundamental Theorem, had enormous power. For eighteenth-century mathematicians, especially those on the Continent where the greatest achievements occurred, it was enough that the concepts of the calculus were understood sufficiently well to be applied to solve a large number of problems, both in mathematics and in physics. So, we come to our third stage: *exploration and development*.

Differential Equations, Taylor Series, and Functions

Newton had stated his three laws of motion in words, and derived his physics from those laws by means of synthetic geometry [15]. Newton's second law stated: *"The change of motion [our 'momentum'] is proportional to the motive force impressed, and is made in the direction of the [straight] line in which that force is impressed"* [15, p. 13]. Once translated into the language of the calculus, this law provided physicists with an instrument of physical discovery of tremendous power—because of the power of the concept of the derivative.

To illustrate, if F is force and x distance (so $m\dot{x}$ is momentum and, for constant mass, $m\ddot{x}$ the rate of change of momentum), then Newton's second law takes the form $F = m\ddot{x}$. Hooke's law of elasticity (when an elastic body is distorted the restoring force is proportional to the distance [in the opposite direction] of the distortion) takes the algebraic form $F = -kx$. By equating these expressions for force, Euler in 1739 could easily both state and solve the differential equation $m\ddot{x} + kx = 0$ which describes the motion of a vibrating spring [10, p. 482]. It was mathematically surprising, and physically interesting, that the solution to that differential equation involves sines and cosines.

An analogous, but considerably more sophisticated problem, was the statement and solution of the partial differential equation for the vibrating string. In modern notation, this is

$$\frac{\partial^2 y}{\partial t^2} = \frac{T\partial^2 y}{\mu \partial x^2},$$

where T is the tension in the string and μ is its mass per unit length. The question of how the solutions to this partial differential equation behaved was investigated by such men as d'Alembert, Daniel Bernoulli, and Leonhard Euler, and led to extensive discussions about the nature of continuity, and to an expansion of the notion of function from formulas to more general dependence relations [10, pp. 502–514], [16, pp. 367–368]. Discussions surrounding the problem of the vibrating string illustrate the unexpected ways that discoveries in mathematics and physics can interact ([16, pp. 351–368] has good selections from the original papers). Numerous other examples could be cited, from the use of infinite-series approximations in celestial mechanics to the dynamics of rigid bodies, to show that by the mid-eighteenth century the differential equation had become the most useful mathematical tool in the history of physics.

Another useful tool was the Taylor series, developed in part to help solve differential equations. In 1715, Brook Taylor, arguing from the properties of finite differences, wrote an equation expressing what we would write as $f(x + h)$ in terms of $f(x)$ and its quotients of differences of various orders. He then let the differences get small, passed to the limit, and gave the formula that still bears his name: the Taylor series. (Actually, James Gregory and Newton had anticipated this discovery, but Taylor's work was more directly influential.) The importance of this property of derivatives was soon recognized, notably by Colin Maclaurin (who has a special case of it named after him), by Euler, and by Joseph-Louis Lagrange. In their hands, the Taylor series became a powerful tool in studying functions and in approximating the solution of equations.

But beyond this, the study of Taylor series provided new insights into the nature of the derivative. In 1755, Euler, in his study of power series, had said that for any power series,

$$a + bx + cx^2 + dx^3 + \dots,$$

one could find x sufficiently small so that if one broke off the series after some particular term— say x^2—the x^2 term would exceed, in absolute value, the sum of the entire remainder of the series

[6, section 122]. Though Euler did not prove this—he must have thought it obvious since he usually worked with series with finite coefficients—he applied it to great advantage. For instance, he could use it to analyze the nature of maxima and minima. Consider, for definiteness, the case of maxima. If $f(x)$ is a relative maximum, then by definition, for small h,

$$f(x - h) < f(x) \text{ and } f(x + h) < f(x).$$

Taylor's theorem gives, for these inequalities,

$$f(x - h) = f(x) - h\frac{df(x)}{dx} + h^2\frac{d^2f(x)}{dx^2} - \dots$$

$$< f(x) \qquad (3)$$

$$f(x + h) = f(x) + h\frac{df(x)}{dx} + h^2\frac{d^2f(x)}{dx^2} + \dots$$

$$< f(x). \qquad (4)$$

Now if h is so small that $h\,df(x)/dx$ dominates the rest of the terms, the only way that both of the inequalities (3) and (4) can be satisfied is for $df(x)/dx$ to be zero. Thus the differential quotient is zero for a relative maximum. Furthermore, Euler argued, since h^2 is always positive, if $d^2f(x)/dx^2 \neq 0$, the only way both inequalities can be satisfied is for $d^2f(x)/dx^2$ to be negative. This is because the h^2 term dominates the rest of the series—unless $d^2f(x)/dx^2$ is itself zero, in which case we must go on and think about even higher-order differential quotients. This analysis, first given and demonstrated geometrically by Maclaurin, was worked out in full analytic detail by Euler [6, sections 253–254], [9, pp. 117–118]. It is typical of Euler's ability to choose computations that produce insight into fundamental concepts. It assumes, of course, that the function in question has a Taylor series, an assumption which Euler made without proof for many functions; it assumes also that the function is uniquely the sum of its Taylor series, which Euler took for granted. Nevertheless, this analysis is a beautiful example of the exploration and development of the concept of the differential quotient of first, second, and nth orders—a development which completely solves the problem of characterizing maxima and minima, a problem which goes back to the Greeks

Lagrange and the Derivative as a Function

Though Euler did a good job analyzing maxima and minima, he brought little further understanding of the nature of the differential quotient. The new importance given to Taylor series meant that one had to be concerned not only about first and second differential quotients, but about differential quotients of any order.

The first person to take these questions seriously was Lagrange. In the 1770s, Lagrange was impressed with what Euler had been able to achieve by Taylor-series manipulations with differential quotients, but Lagrange soon became concerned about the logical inadequacy of all the existing justifications for the calculus. In particular, Lagrange wrote in 1797 that the Newtonian limit-concept was not clear enough to be the foundation for a branch of mathematics. Moreover, in not allowing variables to surpass their limits, Lagrange thought the limit-concept too restrictive. Instead, he said, the calculus should be reduced to algebra, a subject whose foundations in the eighteenth century were generally thought to be sound [11, pp. 15–16].

The algebra Lagrange had in mind was what he called the algebra of infinite series, because Lagrange was convinced that infinite series were part of algebra. Just as arithmetic deals with infinite decimal fractions without ceasing to be arithmetic, Lagrange thought, so algebra deals with infinite algebraic expressions without ceasing to be algebra. Lagrange believed that expanding $f(x + h)$ into a power series in h was always an algebraic process. It is obviously algebraic when one turns $1/(1 - x)$ into a power series by dividing. And Euler had found, by manipulating formulas, infinite power-series expansions for functions like sin x, cos x, e^x. If functions like those have power-series expansions, perhaps everything could be reduced to algebra. Euler, in his book *Introduction to the Analysis of the Infinite (Introductio in analysin infinitorum*, 1748), had studied infinite series, infinite products, and infinite continued fractions by what he thought of as purely algebraic methods. For instance, he converted infinite series into infinite products by treating a series as a very long polynomial. Euler thought that this work was purely algebraic, and—what is crucial here—Lagrange also thought Euler's methods were purely algebraic. So Lagrange tried to make the calculus rigorous by reducing it to the algebra of infinite series.

Lagrange stated in 1797, and thought he had proved, that any function (that is, any analytic expression, finite or infinite) had a power-series expansion:

$$f(x + h) = f(x) + p(x)h + q(x)h^2 + r(x)h^3 + \dots, \quad (5)$$

except, possibly, for a finite number of isolated values of x. He then defined a new function, the coefficient of the linear term in h which is $p(x)$ in the expansion shown in (5) and called it the *first derived function* of $f(x)$. Lagrange's term "derived function" (*fonction dérivée*) is the origin of our term "derivative." Lagrange introduced a new notation, $f'(x)$, for that function. He defined $f''(x)$ to be the first derived function of $f'(x)$, and so on, recursively. Finally, using these definitions, he proved that, in the expansion (5) above, $q(x) = f''(x)/2$, $r(x) = f'''(x)/6$, and so on [11, chapter 2].

What was new about Lagrange's definition? The concept of *function*—whether simply an algebraic expression (possibly infinite) or, more generally, any dependence relation—helps free the concept of derivative from the earlier ill-defined notions. Newton's explanation of a fluxion as a rate of change appeared to involve the concept of motion in mathematics; moreover, a fluxion seemed to be a different kind of object than the flowing

quantity whose fluxion it was. For Leibniz, the differential quotient had been the quotient of vanishingly small differences; the second differential quotient, of even smaller differences. Bishop Berkeley, in his attack on the calculus, had made fun of these earlier concepts, calling vanishing increments "ghosts of departed quantities" [2, section 35]. But since, for Lagrange, the derivative was a function, it was now the same sort of object as the original function. The second derivative is precisely the same sort of object as the first derivative; even the nth derivative is simply another function, defined as the coefficient of h in the Taylor series for $f^{(n-1)}(x + h)$. Lagrange's notation $f'(x)$ was designed precisely to make this point.

We cannot fully accept Lagrange's definition of the derivative, since it assumes that every differentiable function is the sum of a Taylor series and thus has infinitely many derivatives. Nevertheless, that definition led Lagrange to a number of important properties of the derivative. He used his definition together with Euler's criterion for using truncated power series in approximations to give a most useful characterization of the derivative of a function [9, p. 116, pp. 118–121]:

$$f(x + h) = f(x) + hf'(x) + hH, \text{ where } H \text{ goes to zero with } h.$$

(I call this the *Lagrange property of the derivative*.) Lagrange interpreted the phrase "H goes to zero with h" in terms of inequalities. That is, he wrote that,

> Given D, h can be chosen so that $f(x + h)$ $- f(x)$ *lies between* $h(f'(x) - D)$ and $\quad (6)$ $h(f'(x) + D)$.

Formula (6) is recognizably close to the modern delta-epsilon definition of the derivative.

Lagrange used inequality (6) to prove theorems. For instance, he proved that a function with positive derivative on an interval is increasing there, and used that theorem to derive the Lagrange remainder of the Taylor series (9, pp. 122–127], [11, pp. 78–85]. Furthermore, he said, considerations like inequality (6) are what make possible applications of the differential calculus to a whole range of problems in mechanics, in geometry, and, as we have described, the problem of maxima and minima (which Lagrange solved using the Taylor series remainder which bears his name [11, pp. 233–237]).

In Lagrange's 1797 work, then, the derivative is defined by its position in the Taylor series—a strange definition to us. But the derivative is also *described* as satisfying what we recognize as the appropriate delta-epsilon inequality, and Lagrange applied this inequality and its *n*th-order analogue, the Lagrange remainder, to solve problems about tangents, orders of contact between curves, and extrema. Here the derivative was clearly a function, rather than a ratio or a speed.

Still, it is a lot to assume that a function has a Taylor series if one wants to define only *one* derivative. Further, Lagrange was wrong about the algebra of infinite series. As Cauchy pointed out in 1821, the algebra of finite quantities cannot automatically be extended to infinite processes. And as Cauchy also pointed out, manipulating Taylor series is not foolproof. For instance, e^{-1/x^2} has a zero Taylor series about $x = 0$, but the function is not identically zero. For these reasons, Cauchy rejected Lagrange's definition of derivative and substituted his own.

Definitions, Rigor, and Proofs

Now we come to the last stage in our chronological list: *definition*. In 1823, Cauchy defined the derivative of $f(x)$ as the limit, when it exists, of the quotient of differences $(f(x + h) - f(x))/h$ as h goes to zero [4, pp. 22–23]. But Cauchy understood "limit" differently that had his predecessors. Cauchy entirely avoided the question of whether a variable ever reached its limit; he just didn't discuss it. Also, knowing an absolute value when he saw one, Cauchy followed Simon l'Huilier and S.-F. Lacroix in abandoning the restriction that variables never surpass their limits. Finally, though Cauchy, like

Newton and d'Alembert before him, gave his definition of limit in words, Cauchy's understanding of limit (most of the time, at least) was algebraic. By this, I mean that when Cauchy needed a limit property in a proof, he used the algebraic inequality-characterization of limit. Cauchy's proof of the mean value theorem for derivatives illustrates this. First he proved a theorem which states: *if $f(x)$ is continuous on $[x, x + a]$, then*

$$\min_{[x, x+a]} f'(x) \leqslant \frac{f(x + a) - f(x)}{a} \leqslant \max_{[x, x+a]} f'(x). \quad (7)$$

The first step in his proof is [4, p. 44]:

> Let δ, ε be two very small numbers; the first is chosen so that for all [absolute] values of *h* less than δ, and for any value of *x* [on the given interval], the ratio $(f(x + h) - f(x))/h$ will always be greater than $f'(x) - \varepsilon$ and less than $f'(x) + \varepsilon$.

(The notation in this quote is Cauchy's, except that I have substituted *h* for the *i* he used for the increment.) Assuming the intermediate-value theorem for continuous functions, which Cauchy had proved in 1821, the mean-value theorem is an easy corollary of (7) [4, pp. 44–45], [9, pp. 168–170].

Cauchy took the inequality-characterization of the derivative from Lagrange (possibly via an 1806 paper of A.-M. Ampère [9, pp. 127–132]). But Cauchy made that characterization into a definition of derivative. Cauchy also took from Lagrange the name derivative and the notation $f'(x)$, emphasizing the functional nature of the derivative. And, as I have shown in detail elsewhere [9, chapter 5], Cauchy adapted and improved Lagrange's inequality proof-methods to prove results like the mean-value theorem, proof-methods now justified by Cauchy's definition of derivative.

But of course, with the new and more rigorous definition, Cauchy went far beyond Lagrange. For instance, using his concept of limit to define the integral as the limit of sums, Cauchy made a good first approximation to a real proof of the Fundamental Theorem of Calculus [9, pp. 171–175], [4, pp. 122–125, 151–152]. And it was Cauchy who

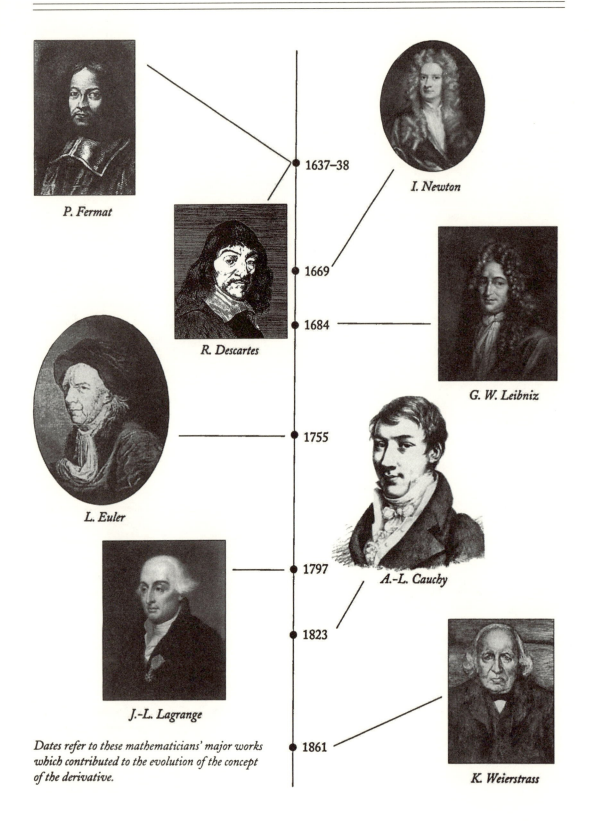

P. Fermat

I. Newton

1637–38

R. Descartes

1669

1684

G. W. Leibniz

L. Euler

1755

A.-L. Cauchy

1797

J.-L. Lagrange

1823

Dates refer to these mathematicians' major works
which contributed to the evolution of the concept
of the derivative.

1861

K. Weierstrass

not only raised the question, but gave the first proof, of the existence of a solution to a differential equation [9, pp. 158–159].

After Cauchy, the calculus itself was viewed differently. It was seen as a rigorous subject, with good definitions and with theorems whose proofs were based on those definitions, rather than merely as a set of powerful methods. Not only did Cauchy's new rigor establish the earlier results on a firm foundation, but it also provided a framework for a wealth of new results, some of which could not even be formulated before Cauchy's work.

Of course, Cauchy did not himself solve all the problems occasioned by his work. In particular, Cauchy's definition of the derivative suffers from one deficiency of which he was unaware. Given an ε, he chose a δ which he assumed would for any x. That is, he assumed that the quotient of differences converged uniformly to its limit. It was not until the 1840s that G. G. Stokes, V. Seidel, K. Weierstrass, and Cauchy himself worked out the distinction between convergence and uniform convergence. After all, in order to make this distinction, one first needs a clear and algebraic understanding of what a limit is—the understanding Cauchy himself had provided.

In the 1850s, Karl Weierstrass began to lecture at the University of Berlin. In his lectures, Weierstrass made algebraic inequalities replace words in theorems in analysis, and used his own clear distinction between pointwise and uniform convergence along with Cauchy's delta-epsilon techniques to present a systematic and thoroughly rigorous treatment of the calculus. Though Weierstrass did not publish his lectures, his students—H. A. Schwartz, G. Mittag-Leffler, E. Heine, S. Pincherle, Sonya Kowalevsky, Georg Cantor, to name a few—disseminated Weierstrassian rigor to the mathematical centers of Europe. Thus although our modern delta-epsilon definition of derivative cannot be quoted from the *works* of Weierstrass, it is in fact the *work* of Weierstrass [3, pp. 284–287]. The rigorous understanding brought to the concept of the derivative by Weierstrass is signaled by his publication in 1872 of an example of an everywhere continuous, nowhere differentiable function. This is a far cry from merely acknowledging that derivatives might not always exist, and the example shows a complete mastery of the concepts of derivative, limit, and existence of limit [3, p. 285].

Historical Development versus Textbook Exposition

The span of time from Fermat to Weierstrass is over two hundred years. How did the concept of derivative develop? Fermat implicity used it; Newton and Liebniz discovered it; Taylor, Euler, Maclaurin developed it; Lagrange named and characterized it; and only at the end of this long period of development did Cauchy and Weierstrass define it. This is certainly a complete reversal of the usual order of textbook exposition in mathematics, where one starts with a definition, then explores some results, and only then suggests applications.

This point is important for the teacher of mathematics: the historical order of development of the derivative is the reverse of the usual order of textbook exposition. Knowing the history helps us as we teach about derivatives. We should put ourselves where mathematicians were before Fermat, and where our beginning students are now—back on the other side, before we had any concept of derivative, and also before we knew the many uses of derivatives. Seeing the historical origins of a concept helps motivate the concept, which we—along with Newton and Leibniz—want for the problems it helps to solve. Knowing the historical order also helps to motivate the rigorous definition—which we, like Cauchy and Weierstrass, want in order to justify the uses of the derivative, and to show precisely when derivatives exist and when they do not. We need to remember that the rigorous definition is often the end, rather than the beginning, of a subject.

The real historical development of mathematics—the order of discovery—reveals the creative mathematician at work, and it is creation that makes doing mathematics so exciting. The order of exposition, on the other hand, is what gives mathematics its characteristic logical structure and its incomparable deductive certainty. Unfortunately, once the classic exposition has been given, the order of discovery is often forgotten. The task of the historian is to recapture the order of discovery: not as we think it might have been, not as we think it should have been, but as it really was. And this is the purpose of the story we have just told of the derivative from Fermat to Weierstrass.

This article is based on a talk delivered at the Conference on the History of Modern Mathematics, Indiana Region of the Mathematical Association of America, Ball State University, April 1982; earlier versions were presented at the Southern California Section of the M. A. A. and at various mathematics colloquia. I thank the *Mathematics Magazine* referees for their helpful suggestions.

REFERENCES

*[1] Margaret Baron, *Origins of the Infinitesimal Calculus*, Pergamon, Oxford, 1969.

[2] George Berkeley, *The Analyst, or a Discourse Addressed to an Infidel Mathematician*, 1734. In A. A. Luce and T. R. Jessop, eds., *The Works of George Berkeley*, Nelson, London, 1951 (some excerpts appear in [16, pp. 333–338]).

[3] Carl Boyer, *History of the Calculus and its Conceptual Development*, Dover, New York, 1959.

[4] A.-L. Cauchy, *Résumé des leçons données a l'école royale polytechnique sur le calcul infinitésimal*, Paris, 1823. In *Oeuvres complètes d'Augustin Cauchy*, Gauthier-Villars, Paris, 1882- , series 2, vol. 4.

[5] Pierre Dugac, *Fondements d'analyse*, in J. Dieudonné, *Abrégé d'histoire des mathématiques, 1700–1900*, 2 vols., Hermann, Paris, 1978.

[6] Leonhard Euler, *Institutiones calculi differentialis*, St. Peterburg, 1755. In *Operia omnia*, Teubner, Leipzig, Berlin, and Zurich, 1911- , series 1, vol. 10.

[7] Pierre Fermat, *Analysis ad refractiones*, 1661. In *Oeuvres de Fermat*, ed., C. Henry and P. Tannery, 4 vols., Paris, 1891–1912; Supplement, ed. C. de Waard, Paris, 1922, vol. 1, pp. 170–172.

[8] ———, *Methodus ad desquirendam maximam et minimam et de tangentibus linearum curvarum*, *Oeuvres*, vol. 1, pp. 133–136. Excerpted in English in [16, pp. 222–225].

[9] Judith V. Grabiner, *The Origins of Cauchy's Rigorous Calculus*, M. I. T. Press, Cambridge and London, 1981.

[10] Morris Kline, *Mathematical Thought from Ancient to Modern Times*, Oxford, New York, 1972.

[11] J.-L. Lagrange, *Théorie des fonctions analytiques*, Paris, 2nd edition, 1813. In *Oeuvres* de Lagrange, ed. M. Serret, Gauthier-Villars, Paris, 1867–1892, vol. 9.

[12] Michael S. Mahoney, *The Mathematical Career of Pierre de Fermat, 1601–1665*, Princeton University Press, Princeton, 1973.

[13] Isaac Newton, *Of Analysis by Equations of an Infinite Number of Terms* [1669], in D. T. Whiteside, ed., *Mathematical Works of Isaac Newton*, Johnson, New York and London, 1964, vol. 1, pp. 3–25.

[14] ———, *Method of Fluxions* [1671], D. T. Whiteside, ed., *Mathematical Works of Isaac Newton*, vol. 1, pp. 29–139.

[15] ———, *Mathematical Principles of Natural Philosophy*, tr. A. Motte, ed. F. Cajori, University of California Press, Berkeley, 1934.

[16] D. J. Struik, *Source Book in Mathematics, 1200–1800*, Harvard University Press, Cambridge, MA, 1969.

[17] D. T. Whiteside, ed., *The Mathematical Papers of Isaac Newton*, Cambridge University Press, 1967–1982.

*Available as a Dover reprint.

2

Niels Henrik Abel

ROBERT W. PRIELIPP

NIELS HENRIK ABEL (1802–1829) is without a doubt the greatest mathematician Norway has ever produced. Though he died before reaching the age of twenty-seven, struck down by tuberculosis, he left a bountiful legacy for mathematics. At the time of his death the world was really just beginning to awaken to his genius. A young man who was well acquainted with poverty, Niels Henrik's family life was not always of the best. Neither his father, who died in 1820, nor his mother was a saint (for a very readable account of Abel's life see [2]), and the family was usually quite heavily in debt. Abel's studies were financed by his professors and by the Norwegian government. He was granted a fellowship to travel and study in Europe, and the year and a half which he spent in Germany, Italy, and France was one of the happiest periods in his life. When Abel returned to Norway, he could only obtain a temporary position. The last years of his life were beset by many difficulties.

Almost every student of beginning algebra is familiar with the quadratic formula. By 1600 Cardan and Ferrari had shown how to solve any cubic equation or quartic equation. The seventeenth and eighteenth centuries witnessed innumerable futile attempts to solve the general equation of fifth degree by radicals. While still in school Abel believed he had found such a solution, but he discovered an error before publication. (At the age

Niels Henrik Abel

of sixteen Galois was to repeat the same mistake, believing for a very short time that he had done what cannot be done. This is just one of the many striking parallels in the careers of Abel and Galois.) Shortly thereafter Abel proved that the equation

$$y^5 - ay^4 + by^3 - cy^2 + dy - e = 0$$

was not solvable by radicals; that is, that y cannot be expressed in terms of a, b, c, d, and e by the use of addition, subtraction, multiplication, and division and extraction of roots only, a finite number

Reprinted from *Mathematics Teacher* 62 (Oct., 1969): 482–84; with permission of the National Council of Teachers of Mathematics.

of times. This paper was published at Oslo in 1824 at Abel's expense. In order to save on printing costs, he had to give the paper in a very sketchy form, which in some places affects the lucidity of his reasoning (for the complete text see [3]). It was this proof which opened the road to the modern theory of equations, including group theory and the solution of equations by means of transcendental functions.

Abel proposed for himself the problem of finding all equations solvable by radicals, and succeeded in solving all equations which had commutative groups (see [1] for an explanation of this terminology). These equations are now called Abelian equations.

Clearly Niels Henrik Abel made a vital contribution to the theory of groups. Listed below are some of his other great achievements.

1. Abel wrote a study of the binomial series, considered to be one of the classics of function theory. It contains the principles of convergent series with special applications to power series.

2. While traveling in Europe, Abel stopped in Berlin. He had the good fortune of making the acquaintance of A. L. Crelle who became his lifelong friend. Inspired by Abel, Crelle founded the important *Journal für die reine und angewandte Mathematik* (*Journal for Pure and Applied Mathematics*), many of the early volumes of which are filled by papers by Abel. These papers concern such topics as equation theory, functional equations, integration in finite form, and problems from theoretical mechanics.

3. During the spring of 1826 Abel completed his *Mémoire sur une propriété générale d'une classe très-étendue de fonctions transcendentes* (*Memoir on a General Property of a Very Extensive Class of Transcendental Functions*), which he himself considered his masterpiece. In this he gives a theory of integrals of algebraic functions—in particular, the result known as Abel's theorem, that there is a finite number, the genus, of independent integrals of this nature. This forms the basis for the later theory of Abelian integrals and Abelian functions. The memoir was submitted to the Paris Academy of Sciences. Legendre and Cauchy were appointed as referees. Somehow Cauchy took the memoir home and mislaid it, and both he and Legendre forgot all about it.

4. The theory of elliptic functions (whose discovery had eluded the great Legendre) was developed with great rapidity in competition with K. G. J. Jacobi.

Two of Abel's greatest recognitions came posthumously. Just two days after Abel's death Crelle wrote to say that Abel would be appointed to the professorship of mathematics at the University of Berlin. Together with Jacobi, Abel was awarded the Grand Prix of the French Academy for 1830.

Perhaps Oystein Ore provided us with the best way to conclude this article when he wrote:

> Mathematicians, in their characteristic manner, have erected many monuments to Abel, more durable than bronze. The custom prevails of marking new results and great ideas by the names of their originators. Today, anyone who reads advanced mathematical texts will find Abel's name perpetuated in numerous branches of his science: Abelian theorems in abundance, Abelian integrals, Abelian equations, Abelian groups, and Abelian formulas. In short, there are few mathematicians whose names are associated with so many concepts of modern mathematics. Niels Henrik would have been greatly surprised at his own importance.

REFERENCES

1. Lieber, Lillian R., and Lieber, Hugh Gray. *Galois and the Theory of Groups: A Bright Star in Mathesis*. Brooklyn: The Galois Institute of Mathematics and Art, 1956.

2. Ore, Oystein. *Niels Henrik Abel, Mathematician Extraordinary*. Minneapolis: University of Minnesota Press, 1957.

3. Smith, David Eugene. *A Source Book in Mathematics*. Vol. I. New York: Dover Publications, 1959, pp. 261–66.

3

Olivier and Abel on Series Convergence: An Episode from Early 19th Century Analysis

MICHAEL GOAR

CALCULUS AND INTRODUCTORY analysis courses are based upon a theoretical framework whose historical development is too often ignored. In reordering and repackaging these theoretical results, authors and instructors often present material in a succession that is natural for an experienced mathematician, but less so for the student. In many cases, the development of a mathematical concept may be better illuminated by studying original source material to *observe* the process of intuitive notions being replaced by more sophisticated ones as the weaknesses of the former are exposed by counterexamples.

Students who encounter this material have the opportunity to witness the difficulties once faced by mathematicians in coming to terms with the meaning of such things as limits and convergence—at the same time as the students' ideas about these concepts are maturing. This historic struggle often parallels students' own difficulties, and enhances their appreciation of their own thinking while deepening their understanding of the underlying mathematics.

We will describe a misstep made while attempting to simplify the process of determining whether an infinite series converges. This episode dates to the 1820s, when the field of analysis was emerging as a formalization of ideas and methods used in calculus, which began its development much earlier. Mathematics historians now regard much of the 19th century as a period of "rigorization" in analysis, where standards of proof, along with a notation we would recognize today, gradually became more widely employed. The misstep came in the form of a simple, universal series convergence test proposed by a mathematician named Louis Olivier, which was subsequently invalidated by the young Norwegian prodigy Niels Abel. The misconception at the heart of this ill-fated conjecture, and the skillfully crafted counterargument forwarded by Abel, make this exchange an excellent study example for students in courses that treat series convergence.

Setting the Stage

We begin with the harmonic series $1 + \frac{1}{2} + \frac{1}{3} + \frac{1}{4} + \cdots$, a standard example that conflicts with students' initial concepts of convergence. Proofs of its divergence vary, but many use a comparison argument put forward by Jakob Bernoulli around 1690, which compares the harmonic series to

Reprinted from *Mathematics Magazine* 72 (Dec., 1999): 347–355; with permission of the Mathematical Association of America and the author.

another positive-termed series which is clearly less, term by term:

$$\sum a_n = 1 + \frac{1}{2} + \frac{1}{3} + \frac{1}{4} + \frac{1}{5} + \frac{1}{6} + \frac{1}{7} + \frac{1}{8} + \frac{1}{9} +$$
$$\dots + \frac{1}{16} + \frac{1}{17} + \dots$$

$$\sum b_n = 1 + \frac{1}{2} + \frac{1}{4} + \frac{1}{4} + \frac{1}{8} + \frac{1}{8} + \frac{1}{8} + \frac{1}{8} + \frac{1}{16} +$$
$$\dots + \frac{1}{16} + \frac{1}{32} + \dots$$

Since Σb_n contains infinitely many groupings of terms that sum to $\frac{1}{2}$, the series diverges. Thus the larger series Σa_n diverges too, by comparison.

A similar proof was given by Cauchy in *Cours d'analyse* (1821), which compiled many of the series convergence tests we recognize in textbooks today, including the root test, the ratio test, the logarithm test, and what is often called Cauchy's condensation test (for a nonnegative, monotone decreasing sequence a_k, $\Sigma_{k=1}^{\infty} a_k$ converges if and only if $\Sigma_{k=0}^{\infty} 2^k a_2^k$ converges). Cauchy writes:

> . . . before effecting the summation of any series, I had to examine in which cases the series can be summed, or, in other words, what are the conditions of their convergence; and on this topic I have established general rules which seem to me to merit some attention.

With so many convergence tests available, each carrying its own set of conditions, one sees a clear motivation for mathematicians then to look for a smaller set of more universal tests, and, in particular, any test that established both necessary and sufficient conditions for convergence. There is a modem parallel for students: After having been convinced in one way or another that the harmonic series truly diverges, and after having seen that the p-series $\Sigma_{n=1}^{\infty} \frac{1}{n^p}$ converges for $p > 1$, students might reasonably conclude that the harmonic series forms a sort

of "boundary" case with which other potentially convergent series of positive terms could be compared. Would a series whose partial sums increase more slowly than those of the harmonic series then necessarily converge?

Just such a claim was made by Louis Olivier in his paper *Remarques sur les series infinies et leur convergence* (Remarks on infinite series and their convergence), published in 1827 in the *Journal für die reine und angewandte Mathematik*, also known as *Crelle's Journal*. Olivier asserts that a series whose terms are positive (or whose terms can be grouped so that each group has a positive sum) and approach zero more rapidly than those of a harmonic series will converge, and vice versa. He states this by proposing a criterion based upon the limit of the product of n and the general term a_n.

Following is an excerpt from Olivier's original paper, and then a translation.

> Therefore if one finds in an infinite series, the product of the nth term, or of the nth group of terms which keep the same sign, by n, is zero, for $n = \infty$, one can regard precisely this situation as an indicator that the series is convergent, and conversely, the series is not convergent if the product $n \cdot a_n$ is nonzero for $n = \infty$.

Notice that it was customary at that time to write "$n = \infty$" where we would write "$n \to \infty$" today. It is interesting to consider how infinite quantities were viewed by those working in this period, a time in which the ε and δ characterizations of limits that are now standard were just emerging in mathematical literature. Olivier continues

> Done si l'on trouve, que dans une série infinie, le produit du n^{me} terms, ou du n^{me} des groupes de termes qui conservent le même signe, par n, est zéro, pour $n = \infty$, on peut regarder cette seule circonstance comme une marque, que la série est convergente; et réciproquement, la série ne peut pas être convergente, sile produit $n.a_n$ n'est pas nul pour $n = \infty$.

Nous allons appliquer ce criterium de la convergence des séries infinies á quelques exemples.

6.

Exemples.

I. Dans la série

$$1 + \frac{1}{2} + \frac{1}{3} + \frac{1}{4} \cdots\cdots + \frac{1}{n} + R.$$

lo n^{me} terme a_n est $\frac{1}{n}$. Done $n \cdot a_n = 1$ n'est pas, 0. Done la série n'est pas convergente.

II. Dans la sériee

$$1 - \frac{1}{2} + \frac{1}{3} - \frac{1}{4} + \frac{1}{5} - \cdots\cdots + \frac{1}{n} - \frac{1}{n+1} + R.$$

ou n est un nombre impair quelconque, le n^{me} des groupes de termes, quiconservent le même signe, est $a_n = \frac{1}{2n-1} - \frac{1}{2n} = \frac{1}{2n(2n-1)}$. Done $n \cdot a_n = \frac{1}{2(2n-1)}$. Ce produit est zéro pour $n = \infty$. Done la série est convergente.

III. Dans la série

$$1 + \frac{1}{2^p} + \frac{1}{3^p} \cdots\cdots + \frac{1}{n^p} + R,$$

le produit $n.a_n$ est égal a $\frac{1}{n^{p-1}}$. Cette quantité est nulle pour $n = \infty$, si $p > 1$. Done la série est convergente, si $p > 1$.

by applying his convergence test to a few examples. Note the use of "R" to indicate the sum of the remaining terms, or "tail," of each series.

We will apply this criterion for the convergence of infinite series to several examples.

I. For the series

$$1 + \frac{1}{2} + \frac{1}{3} + \frac{1}{4} + \dots + \frac{1}{n} + R,$$

the nth term a_n is $\frac{1}{n}$. Thus $n \cdot a_n = 1$ is not 0. Thus the series is not convergent.

In the following example, pairs of terms are grouped, with each group positive, so that the series satisfies Olivier's criterion.

II. For the series

$$1 - \frac{1}{2} + \frac{1}{3} - \frac{1}{4} + \frac{1}{5} - \dots + \frac{1}{n} - \frac{1}{n+1} + R,$$

where n is an arbitrary odd number, the nth group of terms, which keep the same sign, is

$$a_n = \frac{1}{2n-1} - \frac{1}{2n} = \frac{1}{2n(2n-1)}. \text{ Thus }$$

$n \cdot a_n = \frac{1}{2(2n-1)}$. This product is zero for $n = \infty$. Thus the series is convergent.

III. For the series

$$1 + \frac{1}{2^p} + \frac{1}{3^p} \dots + \frac{1}{n^p} + R,$$

the product $n \cdot a_n$ is equal to $\frac{1}{n^{p-1}}$. This quantity is null for $n = \infty$, if $p > 1$. Thus the series is convergent, if $p > 1$.

A Reply from Abel

Before the mid-1800s it was common to see mathematical arguments advanced solely on the weight of examples or by "proofs" that would be better described as appeals to intuition. This state of affairs, in 1826, inspired the young Norwegian mathematician Abel to write:

> I shall devote all my efforts to bring light into the immense obscurity that reigns today in Analysis. It so lacks any plan or system, that one is really astonished that there are so many people who devote themselves to it—and, still worse, it is devoid of any rigor.

Niels Henrik Abel grew up in the small village of Findo, Norway, during a time of great local economic difficulty. His interest in mathematics is credited largely to a remarkable teacher, Bernt Michael Holmboe, who inspired the young Abel to study the works of great mathematicians, including Newton, Euler, and Lagrange. Holmboe patiently guided Abel to assimilate the contents of Gauss's *Disquisitiones Arithmeticae* and to resolve to learn mathematics from original sources produced by such masters. Abel's prodigious success with his studies, complemented by Holmboe's encouragement and financial support, led him to complete his degree at the age of 19 and eventually earn a small stipend to support travel and study in France and Germany, then the center of progress in analysis. In Berlin, Abel met A. L. Crelle, who was publishing the first research journal devoted solely to pure and applied mathematics. This encounter benefited each, as Crelle received excellent papers from Abel, while the young mathematician gained a medium of publication.

Volume 3 of *Crelle's Journal* carried five submissions from Abel, the first entitled *"Note sur le memoire de M. L. Olivier No. 4 du second tome de ce journal, ayant pour titre 'remarques sur les series infinies et leur convergence.'"* (Note on the memoir of Mr. L. Olivier, No, 4 in the second volume of this journal, titled 'Remarks on infinite series and their convergence.') Abel offers a counterexample to Olivier's claim and then proves that *no function of n can be used in the type of limit test for series convergence proposed by Olivier*. Abel's construction is both illuminating and instructive; the following translation is from the original French:

One finds on page 34 of this memoir the following theorem for recognizing if a series is convergent or divergent:

"If one finds in an infinite series, the product of the nth term, or n^{th} group of terms which keep the same sign, by n, is zero, for $n = \infty$, one can regard precisely this situation as an indicator that die series is convergent, and conversely, the series is not convergent if the product $n \cdot a_n$ is nonzero for $n = \infty$."

[1]The French term is "logarithmes hyperbolique," or *hyperbolic logarithms*. The reader may wish to consider why this name was used.

[2]In the original paper, the following line was erroneously printed as

$$\log(1+n) < \frac{1}{n} - \log n = \left(1 + \frac{1}{n \log n}\right) \log n.$$

We assume the sign error is due to the publisher, not to Abel.

The latter part of this theorem is very true, but the first does not appear to be.

For example the series

$$\frac{1}{2\log 2} + \frac{1}{3\log 3} + \frac{1}{4\log 4} + \cdots + \frac{1}{n\log n}$$

is divergent while $n \cdot a_n = \frac{1}{\log n}$ is zero for $n = \infty$.

In fact, the natural logarithms[1] in question are always less than the numbers themselves minus 1, that is, one always has $\log(1 + x) < x$. If $x > 1$,

$$\log(1+x) = x - x^2\left(\frac{1}{2} - \frac{1}{3}x\right) - x^4\left(\frac{1}{4} - \frac{1}{5}x\right)\cdots$$

thus also in this latter case $\log(1 + x) < x$ because $\frac{1}{2} - \frac{1}{3}x, \frac{1}{4} - \frac{1}{5}x,...$ are all positive.

Abel then continues his proof of the divergence of $\Sigma \frac{1}{n\log n}$ by cleverly finding a lower bound for each of its terms.

By letting $x = \frac{1}{n}$ this gives

$$\log\left(1 + \frac{1}{n}\right) < \frac{1}{n} \text{ or, equivalently, } \log\left(\frac{1+n}{n}\right) < \frac{1}{n}$$

or[2]

$$\log(1+n) < \frac{1}{n} + \log n = \left(1 + \frac{1}{n \log n}\right)\log n:$$

thus

$$\log\log(1+n) < \log\log n + \log\left(1 + \frac{1}{n\log n}\right).$$

But since $\log(1 + x) < x$, one has

$$\log\left(1 + \frac{1}{n\log n}\right) < \frac{1}{n\log n} \text{ ; thus, by virtue of}$$

the preceding expression,

$$\log\log(1+n) < \log\log n + \frac{1}{n\log n}.$$

This inequality in hand, Abel sums over all $n > 1$, which, after canceling, leaves the partial sum of his series bounded below by an expression that is unbounded as $n \to \infty$.

By letting successively $n = 2, 3, 4, \ldots$ one finds

$$\log \log 3 < \log \log 2 + \frac{1}{2 \log 2},$$

$$\log \log 4 < \log \log 3 + \frac{1}{3 \log 3},$$

$$\log \log 5 < \log \log 4 + \frac{1}{4 \log 4},$$

$$\cdots$$

$$\log \log (1 + n) < \log \log n + \frac{1}{n \log n}.$$

Therefore, upon taking the sum,

$$\log \log (1 + n) < \log \log 2 + \frac{1}{2 \log 2} + \frac{1}{3 \log 3} +$$

$$\frac{1}{4 \log 4} + \ldots + \frac{1}{n \log n} \qquad .$$

But $\log \log(1 + n) = \infty$ for $n = \infty$; thus the sum of the proposed series

$$\frac{1}{2 \log 2} + \frac{1}{3 \log 3} + \frac{1}{4 \log 4} + \cdots + \frac{1}{n \log n}$$

is infinitely large and consequently the series diverges. The theorem announced in the citation is thus at fault in this case.

After validating his counterexample, Abel extends his argument further, illustrating that a scheme like Olivier's fails not only when taking the limit of $n \cdot a_n$, but in general. He considers the factor n in $n \cdot a_n$ as a specific case of the more general expression $\varphi(n) \cdot a_n$, where the function $\varphi(n)$ (which Abel writes as φn) is used to perform this type of test.

In general, one can demonstrate that it is impossible to find a function φn such that any series

$a_0 + a_1 + a_2 + a_3 + \ldots + a_n$, where we suppose all terms are positive, should be convergent if $\varphi n \cdot a_n$ is zero for $n = \infty$ and divergent in the contrary case. This is what one can make clear with the help of the following theorem.

If the series $a_0 + a_1 + a_2 + a_3 + \cdots + a_n + \cdots$ is divergent, then

$$\frac{a_1}{a_0} + \frac{a_2}{a_0 + a_1} + \frac{a_3}{a_0 + a_1 + a_2} + \cdots + \frac{a_n}{a_0 + a_1 + \cdots + a_{n-1}} + \cdots$$

is so too. In fact, upon remarking that the quantities $a_0 + a_1 + a_2, \ldots$ are positive, one can, by virtue of the theorem $\log(1 + x) < x$, demonstrate the following,

$$\log(a_0 + a_1 + a_2 + \cdots + a_n) - \log(a_0 + a_1 + a_2 + \cdots + a_{n-1}),$$

namely $\log \left(1 + \frac{a_n}{a_0 + a_1 + a_2 + \cdots + a_{n-1}} \right)$, is less than $\frac{a_n}{a_0 + a_1 + a_2 + \cdots + a_{n-1}}$. Thus when taking successively $n = 1, 2, 3 \ldots,$

$$\log(a_0 + a_1) - \log a_0 < \frac{a_1}{a_0},$$

$$\log(a_0 + a_1 + a_2) - \log(a_0 + a_1) < \frac{a_2}{a_0 + a_1},$$

$$\log(a_0 + a_0 + a_1 + a_3) - \log(a_0 + a_1 + a_2) < \frac{a_3}{a_0 + a_1 + a_2},$$

$$\cdots$$

$$\log(a_0 + a_1 + \cdots + a_n) - \log(a_0 + a_1 + \cdots + a_{n-1}) < \frac{a_n}{a_0 + a_1 + \cdots + a_{n-1}},$$

and, upon taking the sum,

$$\log(a_0 + a_1 + \cdots + a_n) - \log a_0 < \frac{a_1}{a_0} + \frac{a_2}{a_0 + a_1} + \cdots + \frac{a_3}{a_0 + a_1 + \cdots + a_{n-1}}.$$

But if the series $a_0 + a_1 + \cdots + a_n$ is divergent, its sum is infinite, and the logarithm of this sum is so likewise, so the sum of the series

$$\frac{a_1}{a_0} + \frac{a_2}{a_0 + a_1} + \cdots + \frac{a_n}{a_0 + a_1 + \cdots + a_{n-1}}$$

is also infinitely large, and this series is consequently divergent if the series $a_0 + a_1 + a_2 + \cdots + a_{n-1}$ is.

Note the similarity between this argument and the one Abel uses to show the divergence of $\Sigma \frac{1}{n \log n}$. Finally, Abel uses the theorem he has just demonstrated (if a series $a_0 + a_1 + a_2 + a_3 + \cdots + a_n + \cdots$ diverges, then

$$\frac{a_1}{a_0} + \frac{a_2}{a_0 + a_1} + \frac{a_3}{a_0 + a_1 + a_2} +$$

$$\cdots + \frac{a_3}{a_0 + a_1 + \cdots + a_{n-1}} + \cdots$$

also diverges) to show that any series convergence test like Olivier's will lead to a contradiction.

This now stated, we suppose that φn be a function of n such that the series $a_0 + a_1 + a_2 + \cdots + a_n + \cdots$ is convergent or divergent according as $\varphi n \cdot a_n$ is zero or not for $n = \infty$. Then the series

$$\frac{1}{\varphi 1} + \frac{1}{\varphi 2} + \frac{1}{\varphi 3} + \frac{1}{\varphi 4} + \cdots + \frac{1}{\varphi n} + \cdots$$

will be divergent, and the series

$$\frac{1}{\varphi 2 \cdot \dfrac{1}{\varphi 1}} + \frac{1}{\varphi 3 \left(\dfrac{1}{\varphi 1} + \dfrac{1}{\varphi 2} \right)} + \frac{1}{\varphi 4 \left(\dfrac{1}{\varphi 1} + \dfrac{1}{\varphi 2} + \dfrac{1}{\varphi 3} \right)} + \cdots$$

$$+ \frac{1}{\varphi n \left(\dfrac{1}{\varphi 1} + \dfrac{1}{\varphi 2} + \dfrac{1}{\varphi 3} + \cdots + \dfrac{1}{\varphi(n-1)} + \cdots \right)}$$

convergent; because in the first, one has $a_n \cdot \varphi n = 1$, and in the second, $a_n \cdot \varphi n = 0$ for $n = \infty$.

(The reader might consider why, in the second series, $a_n \cdot \varphi(n) \to 0$ as $n \to \infty$, as Abel claims.) At the end we see Abel apply his argument to Olivier's criterion (in which case $\varphi (n) = n$), encountering the same contradiction as he does in the general case.

Now following the theorem established before, the second series is necessarily divergent, whenever the first is, thus a function φn such as assumed here, doesn't exist. By taking $\varphi n = n$, the two series in question become

$$1 + \frac{1}{2} + \frac{1}{3} + \frac{1}{4} + \cdots + \frac{1}{n} + \cdots$$

and

$$\frac{1}{2 \cdot 1} + \frac{1}{3 \left(1 + \dfrac{1}{2}\right)} + \frac{1}{4 \left(1 + \dfrac{1}{2} + \dfrac{1}{3}\right)} +$$

$$\cdots + \frac{1}{n \left(1 + \dfrac{1}{2} + \dfrac{1}{3} + \cdots + \dfrac{1}{n-1}\right)} + \cdots$$

which consequently are both divergent.

Conclusion

Counterexamples and proof by contradiction are standard tools for distinguishing necessary from sufficient conditions in deductive reasoning. Here we see a beautiful piece of mathematics employing both to clarify a misconception. It is interesting from a historical perspective to note that Abel proves a version of what we now call the Abel-Dini theorem, established in its general form in 1867 by U. Dini.

THEOREM. *If $\Sigma_{n=1}^{\infty} d_n$ is an arbitrary divergent series of positive terms, and*

$$D_n = d_1 + d_2 + \cdots + d_n$$

are its partial sums, the series

$$\sum_{n=1}^{\infty} a_n \equiv \sum_{n=1}^{\infty} \frac{d_n}{D_n^{\alpha}}$$

converges when $\alpha > 1$ and diverges when $\alpha \leq 1$.

In his reply to Olivier, Abel proves a result similar to the case where $\alpha = 1$.

Although $\lim_{n \to \infty} n \cdot a_n = 0$ does not imply convergence of Σa_n, as claimed by Olivier in 1827, a partial converse holds for a series of positive, monotone decreasing terms:

$$\sum a_n \text{ converges} \Rightarrow \lim_{n \to \infty} n a_n = 0.$$

This is what Abel probably meant by remarking that "the latter part of this theorem is very true…".

Is there a universal test for convergence? Does a series exist that converges or diverges more slowly than any other? These and other fundamental questions about the convergence of positive-termed series were ultimately resolved by the close of the 19th century, combining work of many mathematicians.

Acknowledgment. Source material for this paper was drawn from a set of notes and readings prepared by Dr. Otto Bekken of Agder Högskolan, in Kristiansand, Norway. I would like to thank Dr. Bekken for sharing his love for and knowledge of mathematics history, and for his help and inspiration toward this work. I would also like to thank Dr. David Pengelley at New Mexico State University for his efforts to motivate current and future teachers to prepare teaching materials based upon original sources, for his help with translating, and for his patience in reviewing this paper.

4

The Parallel Postulate

RAYMOND H. ROLWING
AND MAITA LEVINE

$\mathcal{E}$UCLID'S FAMOUS PARALLEL postulate was responsible for an enormous amount of mathematical activity over a period of more than twenty centuries. The failure of mathematicians to prove Euclid's statement from his other postulates contributed to Euclid's fame and eventually led to the invention of non-Euclidean geometries.

Before Euclid's time, various definitions of parallel lines had been considered by the Greeks and then discarded. Among them were "parallel lines are lines everywhere equidistant from one another" and "parallel lines are lines having the same direction from a given line." But these early definitions were sometimes vague or contradictory. Euclid tried to overcome these difficulties by his definition, "Parallel lines are straight lines which, being in the same plane and being produced indefinitely in both directions, do not meet one another in either direction," and by his fifth postulate, "Let it be postulated that, if a straight line falling on two straight lines makes the interior angles on the same side less than two right angles, the two straight lines, if produced indefinitely, meet on that side on which the angles are less than two right angles."

The statements of Euclid's first four assumptions are: "Let the following be postulated: (1) To draw a straight line from any point to any point. (2) To produce a finite straight line continuously in a straight line. (3) To describe a circle with any center and distance. (4) That all right angles are

Reprinted from *Mathematics Teacher* 62 (Dec., 1969): 665-69; with permission of the National Council of Teachers of Mathematics.

equal to one another." All of his assumptions fall into one of two categories. The first is the set of "self-evident" facts concerning plane figures. An example of such an assumption is that "a straight line is the shortest distance between two points." The second category deals with concepts beyond the realm of actual experience. For example, Euclid stated that "a straight line must continue undeviatingly in either direction without end and without finite length." Since it is impossible to experience things indefinitely far off, anything that is said about events there is speculation, not self-evident truth. The fifth postulate falls into this latter category.

The complicated nature of the fifth postulate led numerous mathematicians to believe that it could be proved using the remaining postulates, and, therefore, ought to be a theorem rather than a postulate. Even Euclid might have supported this viewpoint since he did succeed in proving the converse of the postulate. One of the first geometers who attempted to prove a statement equivalent to Euclid's parallel postulate was Posidonius, in the first century B.C. He had defined parallel lines as lines that are coplanar and equidistant. A second early important attempt to prove the parallel postulate was made by Claudius Ptolemy of Alexandria, in the second century.

In the fifth century, Proclus, who had studied mathematics in Alexandria and taught in Athens, worked extensively on the problem of proving Euclid's fifth postulate. He succeeded in showing that the postulate could be proved if the following statement could be established: If L_1 and L_2 are any two parallel lines and L_3 any line distinct from and intersecting L_1, then L_3 intersects L_2.

In his argument Proclus used the phrase "distance between parallels." Thus, he assumed that parallel lines are everywhere equidistant, and this assumption is logically equivalent to the fifth postulate. In effect, Proclus assumed what he was trying to prove.

The most elaborate attempt to prove the parallel postulate, and the one most significant for the further development of geometry, was made by the Italian priest, Girolamo Saccheri, who taught mathematics at the University of Pavia. His significant work, published in 1733, was entitled *Euclides ab omni naevo vindicatus sive conatus geometricus quo stabiliuntur prima ipsa geometriae principia*. In this treatise, Saccheri tried to free Euclid from all error, including the supposed error of assuming Postulate V.

One of Saccheri's contributions was the introduction of the Saccheri quadrilateral. The construction of this quadrilateral is as follows. At the endpoints of a segment $\overline{AB}$ construct congruent segments $\overline{AC}$ and $\overline{BD}$, each perpendicular to $\overline{AB}$ and draw $\overline{CD}$. Saccheri tried to prove, on the basis of the first four postulates, that $m \triangle ACD = m \triangle BCD$. He reasoned that if P is the midpoint of $\overline{AB}$ and if Q is the midpoint of $\overline{CD}$ then rt. $\triangle CAP \cong$ rt. $\triangle DBP$, whence $m \angle ACP = m \angle BDP$ and $CP = DP$. Then $\triangle CPQ \cong \triangle DPQ$, so $m \angle PCQ = m \angle PDQ$, and, therefore, $m \angle ACD = m \angle BCD$. This proof does not depend upon the parallel postulate. Saccheri called $\angle ACD$ and $\angle BDC$ the *summit angles* of the quadrilateral and he formu-

lated the following three possibilities, which are exhaustive and pairwise mutually exclusive: (1) The summit angles are right angles. (2) The summit angles are obtuse angles. (3) The summit angles are acute angles. These possibilities are generally called the *right angle hypothesis*, the *obtuse angle hypothesis*, and the *acute angle hypothesis*, respectively. Saccheri succeeded in proving that if any of these hypotheses is valid for one Saccheri quadrilateral, it is valid for every quadrilateral of the same type. He proved further that the fifth postulate is a consequence of the right angle hypothesis. And, by asssuming that a straight line is infinitely long, he showed that the obtuse angle hypothesis is self-contradictory.

Disposing of the acute angle hypothesis presented some difficulty, so Saccheri argued intuitively that the "hypothesis of the acute angle is absolutely false, because it is repugnant to the nature of a straight line." Actually, no logical contradiction can be deduced from the acute angle hypothesis, for it gives rise to a new geometry.

So, while Saccheri was looking for a proof of the parallel postulate, he discovered a new world, the world of "absolute geometry," whose theory is independent of the question of the parallel postulate. Included in "absolute geometry" are the theorems concerning congruent triangles, the inequalities involving the measures of the sides and angles of a triangle, and a set of theorems about the Saccheri quadrilateral. The quadrilateral $ABCD$ is a Saccheri quadrilateral if $\angle A$ and $\angle B$ are right angles and $AC = BD$. Among the theorems that can be proved are: (1) The diagonals of a Saccheri quadrilateral are always congruent. (2) In any Saccheri quadrilateral, the upper base angles are congruent, i.e., $\angle C \cong \angle D$. (3) In any Saccheri quadrilateral, the upper base is congruent to or longer than the lower base, i.e., $CD \geq AB$.

After Saccheri's time, many mathematicians pursued the problem of trying to prove the parallel postulate from the first four postulates. In 1766, J. H. Lambert, a Swiss geometer, showed that Saccheri's obtuse angle hypothesis is consistent

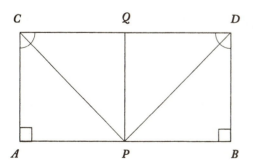

THE SEARCH FOR CERTAINTY

with spherical geometry. In many cases those who attacked the problem worked with statements that are logically equivalent to the fifth postulate rather than with the statement of Euclid. Legendre (1752–1833) tried to prove the following alternative to Euclid's postulate: There exists a triangle in which the sum of the measures of the three angles is equal to the sum of the measures of two right angles. He presented a proof of the fact that the sum of the measures of the angles of a triangle cannot be greater that 180°, but he failed to supply a proof of the fact that the sum cannot be less than 180°. At any rate, his alleged proof of the latter rests on assumptions which are equivalent to the theorem he was trying to establish.

In 1809, Bernhard Friedrich Thibaut tried to demonstrate the existence of a triangle with the property that the sum of the measures of the angles is equal to 180°. His argument was based on the assumption that every rigid motion can be resolved into a rotation and an independent translation. This assumption, however, is equivalent to Postulate V.

In 1813, John Playfair copied Thibaut's argument and tried to correct Thibaut's errors. His attempt was unsuccessful, but it is his alternate statement of the fifth postulate that is now best known and most frequently quoted: "Through a given point, not on a given line, only one parallel can be drawn to the given line." Playfair's statement is surely simpler and more direct than Euclid's.

Karl Friedrich Gauss of Göttingen, Germany, studied the theory of parallels for thirty years. His work resulted in the formulation of a non-Euclidean geometry. In a letter to his friend Franz Adolf Taurinus, dated November 8, 1824, he stated: "The assumption that the angle sum [of a triangle] is less that 180° leads to a curious geometry, quite different from ours but thoroughly consistent, which I have developed to my entire satisfaction. The theorems of this geometry appear to be paradoxical, and, to the uninitiated, absurd, but calm, steady reflection reveals that they contain nothing at all impossible." However, Gauss wrote only a short account of his geometry, which was pub-

lished in 1831. He apparently shrank from the controversy in which a treatise on the new geometry would have involved him. Along with other eminent mathematicians he was influenced by the authority of the German philosopher Immanuel Kant, who had died in 1804. Kant's doctrine emphasized that Euclid's geometry was "inherent in nature." Although Plato had said merely that God geometrizes, Kant asserted, in effect, that God geometrizes according to Euclid's *Elements*.

Gauss was the forerunner of the group of geometers who, instead of trying to prove Euclid's parallel postulate, replaced it by a contradiction of it and thereby invented a new geometry. In 1823, Janos Bolyai replaced Postulate V with the statement: "In a plane two lines can be drawn through a point parallel to a given line and through this point an infinite number of lines may be drawn lying in the angle between the first two and having the property that they will not intersect the given line." Bolyai's first disappointment came when he learned that Gauss had worked on the same problem for thirty years and had achieved the same results. He was disappointed a second time when he discovered that Nikolai Ivanovich Lobachevsky, a professor at the University of Kasan in Russia, had also invented the new geometry and published an account of it in 1829. Bolyai's work was published in 1833 as a twenty-six-page appendix to a semiphilosophical two-volume treatise on elementary mathematics written by his father.

Lobachevsky is now accorded most of the credit for the invention of the new geometry, and his name is the one usually attached to it. His replacement of Euclid's fifth postulate was stated as follows: "Through a point P not on a line there is more than one line which is parallel to the given line." The rest of Euclid's postulates were preserved. A frequently used model of Lobachevskian geometry is the Poincaré model. A detailed description of this model can be found in *Elementary Geometry from an Advanced Standpoint*, by Moise. Of course, any theorems of Euclid's geometry which do not depend on the parallel postulate are valid in

Lobachevsky's geometry. On the other hand, the following theorems are examples of statements which are quite different from the corresponding theorems of Euclidean geometry.

1. *No quadrilateral is a rectangle; if a quadrilateral has three right angles, the fourth angle is acute.*
2. *The sum of the measures of the angles of a triangle is always less than 180°.*
3. *If two triangles are similar, they are congruent.*

Thus, through the discoveries of Gauss, Bolyai, and Lobachevsky, mathematicians recognized the existence of more than one consistent geometry. Leonard M. Blumenthal summarized the significance of the work of these three geometers:

> When the culture was ripe for it, three men, Gauss, Bolyai, and Lobachevsky (a German, a Hungarian, and a Russian) arose in widely separated parts of the learned world, and working independently of one another, created a new geometry. It would be difficult to overestimate the importance of their work. A significant milestone in the intellectual progress of mankind had been passed.

The first period in the history of non-Euclidean geometry, in the opinion of Felix Klein, ended with Lobachevsky's research. This period was characterized by the use of synthetic methods. Riemann, Helmholtz, Lie, and Beltrami were the representatives of the second period in the history of non-Euclidean geometry. Their work involved using the tools of differential geometry.

G. F. B. Riemann is credited with the development of another non-Euclidean geometry, in 1854, which can be realized on a sphere. He began by studying Euclid's postulate that a straight line has infinite length. Discarding this assumption, he invented a geometry in which all lines have finite length. Euclid's first postulate was replaced by the statement: "A straight line is restricted in length and without endpoints." And the parallel postulate was replaced by the statement: "through a point in a plane there can be drawn in the plane no line which does not intersect a given line not passing through the given point." The remainder of Euclid's postulates were retained. In Riemannian geometry, the following theorems can be proved:

1. *Two perpendiculars to the same line intersect.*
2. *Two lines enclose an area.*
3. *The sum of the measures of the angles of a triangle is greater that 180°.*
4. *If two sides of a quadrilateral are congruent and perpendicular to a third side, the figure is not a rectangle, since two of the angles are obtuse.*

An important contribution to the study of non-Euclidean geometry was made by the Italian mathematician, Eugenio Beltrami. His paper, published in 1868, gave the final answer to the question of the consistency of the new geometries. Bolyai and Lobachevsky had suspected that extending their investigations to three-dimensional space might reveal inconsistencies. Beltrami's paper gave an interpretation of plane non-Euclidean geometry as the geometry of geodesics on a certain class of surfaces in Euclidean space. Therefore, the new geometries must be as consistent as Euclidean geometry.

Gauss was the first to use the term "non-Euclidean" geometry, and Felix Klein first gave to the new geometries the names currently used to describe them. He called Lobachevsky's geometry *hyperbolic*, Riemann's geometry *elliptic*, and Euclid's geometry *parabolic*. This terminology arose from the projective approach to non-Euclidean geometry developed by Klein and Arthur Cayley.

Non-Euclidean geometry not only widened the scope of geometric knowledge; it also stimulated discussions concerning what is a geometry and what is "mathematical truth." As recently as the 1820s, geometry had been thought of as an idealized description of the spatial relations of the world in which we live. Euclid held this viewpoint on the meaning of geometry, and chose for his postulates statements that had their roots in everyday experience. A geometric statement was then regarded as true if it correctly described nature, and false if it did not. But Beltrami's proof that Euclid's geometry was not the only consistent geometry forced mathematicians to abandon the idea that geometric truth involved a description of nature.

David Hilbert called the invention of non-Euclidean geometry "the most suggestive and notable achievement of the last century." And Heath claimed that one of the cornerstones on which Euclid's greatness as a mathematician rests was his parallel postulate, for "when we consider the countless successive attempts made through more than twenty centuries to prove the postulate, many of them by geometers of ability, we cannot but admire the genius of the man who concluded that such a hyopothesis, which he found necessary to the validity of his whole system of geometry, was really undemonstrable."

BIBLIOGRAPHY

Archibald, Raymond Clare. "Outline of the History of Mathematics, Part II," *American Mathematical Monthly*, LVI (January 1949), 21–26.

Bell, E. T. *The Development of Mathematics*. 1st ed. New York: McGraw-Hill Book Co., 1940. Pp. 304–7.

* Blumenthal, Leonard M. *A Modern View of Geometry*. San Francisco: W. H. Freeman & Co., 1961. Pp. 4–17.

Cajori, Florian. *A History of Mathematics*, 2d ed. New York: Macmillan Co., 1919. Pp. 48, 306.

Eves, Howard, and Newsom, Carroll V. *An Introduction to the Foundations and Fundamental Concepts of Mathematics*. Rev. ed. New York: Holt, Rinehart & Winston, 1965. Pp. 58–79.

Merriman, Gaylord M. *To Discover Mathematics*. New York: John Wiley & Sons, 1942. Pp. 144–51.

Moise, Edwin. *Elementary Geometry from an Advanced Standpoint*. Reading, Mass.: Addison-Wesley Publishing Co., 1963. Pp. 115–31.

Sanford, Vera. *A Short History of Mathematics*. Boston: Houghton Mifflin Co., 1930. Pp. 276–81.

*Available as a Dover reprint.

5

Saccheri, Forerunner of Non-Euclidean Geometry

SISTER MARY OF MERCY FITZPATRICK

*F*EW PEOPLE CONTEST the statement that Lenin, the father of the Russian Revolution, effected the most drastic social upheaval of our modern times. Nor do they argue whether or not the monuments erected or the volumes published to commemorate this feat are in proportion to the enormity of the achievement. These facts are accepted without question or comment. Might it not occur to us to search for a foundation for the old saying that "history repeats itself," in this particular case? And if a search were made, would we not all be astonished to find that no parallel seems to be evident? For just a century earlier, Lenin's countryman, Nikolai Ivanovich Lobachevski (1793–1856), who had brought about the most famous mathematical revolution of all times, received no honors, but was, instead, removed from his post as rector and professor of mathematics at the University of Kazan. No explanation whatsoever was given to this dedicated mathematician who had served the university well, and who according to Edna E. Kramer in her book, *The Main Stream of Mathematics*, was becoming renowned throughout Europe.

Today, Lobachevski's achievement, the discovery of a new and logical non-Euclidean geometry, is universally hailed, not only by the mathematicians, but also by laymen as well. But what, perhaps, is not so well known is the fact that, just as in the case of calculus the propitious mo-

Reprinted from *Mathematics Teacher* 57: (May, 1964): 323–32; with permission of the National Council of Teachers of Mathematics.

ment for discovery was the age of Newton and Leibniz, so in the matter of Lobachevskian geometry the way was clearly well paved by scores of mathematicians who had attempted to prove Euclid's Fifth Postulate. One of the most notable of these mathematicians, chiefly because of the sincerity of his work and his proven ability as a logician, was a man named Girolamo Giovanni Saccheri (1667–1733), with whom this paper is concerned.

Girolamo Giovanni Saccheri was born at San Remo, Italy, on the night of September 4, 1667. From his earliest years he showed extreme precociousness and a spirit of inquiry, but little else is known of his childhood and early adolescence. In March, 1685, he entered the Order of the Society of Jesus in Genoa. Having completed his novitiate in 1690, he was sent by his superiors to the Collegio di Brera in Milan to teach grammar and, at the same time, to study theology and philosophy. It was here that he met the famous professor of mathematics, the Jesuit Father Tomasso Ceva, for whose brother, Giovanni, the famous Ceva theorem of college geometry is named. Here, too, he became acquainted with the important mathematicians of his day, among them Viviani, who called himself the last pupil of Galileo.

Father Ceva introduced the young scholastic, Saccheri, to the reading of Euclid's *Elements* in the edition of Clavius in which the Fifth Postulate of Euclid is listed as Axiom 13. This assignment apparently became a lifetime pursuit. It seems rather strange, however, that Saccheri at this time seems to have paid little, if any, attention to the contemporary discoveries of Newton and Leibniz, although he had analyzed the work of another contemporary, John Wallis (1616–1703), and had pointed out flaws in the latter's proof of the Fifth Postulate.

In 1694, Saccheri was sent to teach philosophy and polemic theology in the Collegio dei Gesuiti of Turin, at which post he remained for a period of three years. As a result of his studies and teaching during the three years, he produced a little book, *Logica demonstrativa*, which, according to his biographer, Alberto Pascal, deserves to be better known than it is by mathematicians and logicians.

The first edition of the *Logica demonstrativa*, which appeared in 1697, did not carry the author's name, but was published under the cover of a thesis by Count Gravere, one of Saccheri's students. There has been much speculation about why Saccheri did not allow his name to appear at this time. Some think that it was because of his standing; he had not yet attained the rank of professor, and as a result, even though his genius was evident, the work would not receive the acclaim it deserved. Others say that his religious superiors did not permit him to publish the work under his own name. Whatever the reason, the book reappeared four years later when, armed with the rank of professor and a change of residence, Saccheri's reputation was soaring in university circles. He was now acclaimed as a brilliant teacher and, incidentally, a man of such remarkable memory that, according to Vera Sanford in *A Short History of Mathematics*, he could play three games of chess at the same time without seeing any of the boards.

A third edition of the *Logica demonstrativa* appeared in 1735, two years after Saccheri's death. Were it not for Giovanni Vailati, who in 1903 brought this work to the attention of the world in his own small publication, it might well have been cast into oblivion. In the *Logica demonstrativa*, the only existing copy of which is preserved in the Stadtbibliothek of Cologne, Saccheri lays down the clear distinction between what he calls

definitiones quid nominis and *definitiones quid rei*, or between *nominal* and *real* definitions. The former are only intended to explain the meaning that is to be attached to a given term, whereas the latter, besides declaring the meaning of a word, affirm at the same time the existence of the thing defined or, in geometry, the possibility of constructing it. The nominal definition becomes a real definition by means of a *postulate* or an affirmative answer to the question of whether or not the thing exists. In this work Saccheri showed his absorption in the powerful method of *reductio ad absurdum*, an idea he later applied to an investigation of Euclid.

When Saccheri, a full-fledged member of the Society of Jesus, was appointed professor of mathematics at the University of Pavia, he joined the ranks of those mathematicians who had already set out to discover flaws in Euclid's fundamental postulates. The well-known postulate of parallel lines, was not, it was felt, sufficiently obvious. The Greek followers of Euclid had made attempts to prove it; the Arabians, when they acquired the Greek mathematics, also found the parallel axiom unsatisfactory. No one doubted that this was a necessary truth, but many felt that there should be some way of deducing it from the other and simpler axioms of Euclid.

As almost every high school student of geometry knows today, Euclid organized into a deductive system the fundamental geometry known in his day. He selected a few simple geometric facts as a basis and sought to demonstrate that all remaining facts were logical consequences of these. For his basic facts, which he called axioms or postulates, he gave no proof, but used them as the foundation of his system. The five geometric statements selected by Euclid as the basis of his deductive treatment of geometry are introduced as follows:

Let the following be postulated:

1. To draw a straight line from any point to any point.

2. To produce a finite straight line.

3. To draw a circle with any center and distance.

4. That all right angles are equal to one another.

5. That if a straight line falling on two straight lines makes the interior angles on the same side less than two right angles, the two straight lines if produced indefinitely meet on that side on which are the angles less than two right angles.

It is largely to this last (fifth) postulate that Euclid owes his greatness, and yet it is the one which has been the basis of the sharpest attacks on his system. The other four postulates are simple and in sharp contrast to the complicated nature of the statement of the Fifth Postulate. It is not surprising, then, that mathematicians even in Euclid's own time suggested that it should be a theorem rather than a postulate and initiated the efforts to demonstrate it as a logical consequence of the first four. To review these efforts is beyond the scope of this paper, since it is Saccheri's work that claims our attention.

Saccheri, as was stated earlier, had developed an interest in the method of *reductio ad absurdum.* Hence he set out to develop the consequences of denying Euclid's parallel axiom while retaining all the others. In this way he expected to develop a geometry which should be self-contradictory, since he had no doubt that the parallel axiom was a necessary truth.

To arrive at a contradiction in the most convenient form, Saccheri employed a figure found in his Clavius (1574), the birectangular quadrilateral. In his "Indicis loco" of *Euclides vindicatus* he states:

1. In Propositions I and II of the First Book two principles are established, from which in Propositions III and IV is proved that interior angles at the straight joining the extremities of equal perpendiculars erected toward the same parts (in the same plane) from two points of another straight, as base, not merely are equal to each other, but besides are either right or obtuse or acute according as that join is equal to, or less, or greater than the aforesaid base; and inversely.

2. Hence occasion is taken to distinguish three different hypotheses, one of right angle, another of

obtuse, a third of acute: about which, in Propositions V, VI, and VII, is proved that any of these hypotheses is always alone true if it is found true in any one particular case.

Let us examine Propositions I, II, III, and IV. Proposition I states:

If two equal straights (Figure 1) *AC, BD* make with the straight *AB* angles equal toward the same parts: I say that the angles at the join *CD* will be mutually equal.

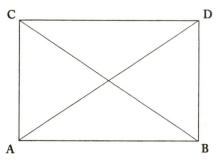

FIGURE I

Saccheri's proof is very simple:

Join *AD, CB.* Then consider the triangles *CAB, DBA.* It follows (Eu. I 4) that the bases *CB, AD* will be equal. Then consider the triangles *ACD, BDC.* It follows (Eu. I 8) that the angles *ACD, BDC* will be equal.

Q.E.D.

Proposition II states:

Retaining the uniform quadrilateral *ABCD*, bisect the sides *AB, CD* (Figure 2) in the points *M* and *H*. I say the angles at the join *MH* will then be right.

Proof: Join *AH, BH,* and likewise *CM, DM.* Because in this quadrilateral the angles *A* and *B* are taken equal and likewise (from the preceding proposition) the angles *C* and *D* are equal; it follows (Eu. I 4, noting the equality of the sides) that in the triangles *CAM, DBM,* the bases *CM, DM* will be equal; and likewise, in the triangles *ACH, BDH,* the bases *AH* and *BH*. Therefore, comparing the triangles *CHM, DHM,* and in turn the triangles *AMH, BMH,* it follows (Eu. I 8) that we have mutually equal, and therefore right, the angles at the points *M* and *H*.

Q.E.D.

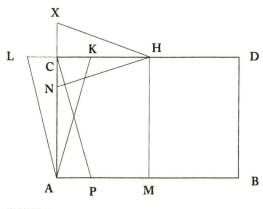

FIGURE 3

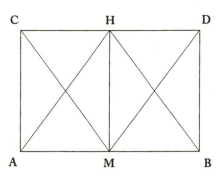

FIGURE 2

Proposition III states:

If two equal straights (Figure 3) *AC, BD,* stand perpendicular to any straight *AB:* I say the join *CD* will be equal to, or less than, or greater than *AB,* according as the angles at *CD* are right, or obtuse, or acute.

Saccheri divides this proof into three parts as follows:

1. Each angle *C* and *D* being right; suppose, if it were possible, either one of these, as *DC,* greater than the other *BA.*

Take in *DC* the piece *DK* equal to *BA,* and join *AK.* Since therefore on *BD* stand perpendicular the equal straights *BA, DK,* the angles *BAK* and *DKA* will be equal (Prop. I). But this is absurd; since the angle *BAK* is by construction less than the assumed right angle *BAC;* and the angle *DKA* is by construction external, and therefore (Eu. I 16) greater than the internal and opposite *DCA* which is supposed right. Therefore neither of the aforesaid straights *DC, BA,* is greater than the other, whilst the angles at the join *CD* are right; and therefore they are mutually equal.

Q.E.D. *(primo loco)*

2. But if the angles at the join *CD* are obtuse, bisect *AB* and *CD* in the points *M* and *H,* and join *MH.*

Since therefore on the straight *MH* stand perpendicular (Prop. II) the two straights *AM, CH,* and at the join *AC* is a right angle at *A,* the straight *CH* (Prop. I) will not be equal to this *AM,* since a right angle is lacking at *C.*

But neither will it be greater: otherwise in *HC* the piece *KH* being assumed equal to this *AM,* the angles at the join *AK* will be (Prop. I) equal.

But this is absurd, as above. For the angle *MAK* is less than a right: and the angle *HKA* is (Eu. I 16) greater than an obtuse, such as the internal and opposite *HCA* is supposed.

It remains therefore, that *CH,* whilst the angles at the join *CD* are taken obtuse, is less than this *AM;* and therefore *CD,* double the former, is less than *AB,* double the latter.

Q.E.D. *(secundo loco)*

3. Finally, however, if the angles at the join are acute, *MH* being constructed perpendicular as before (Prop. II), we proceed thus. Since on the straight *MH* stand perpendicular two straights *AM, CH,* and at the join *AC* is a right angle at *A,* the straight *CH* will not be equal to this *AM* (as above), since the angle at *C* is not right. But neither will it be less otherwise, if in *HC* produced *HL* is taken equal to this *AM,* the angles at the join *AL* will be (as above) equal.

But this is absurd. For the angle *MAL* is by construction greater than the assumed right *MAC;* and the angle *HLA* is by construction internal, and opposite, and therefore less than (Eu. I 16) the external *HCA,* which is assumed acute.

It remains, therefore, that *CH*, whilst the angles at the join *CD* are acute, is greater than this *AM*, and therefore *CD*, the double of the former, is greater than *AB*, the double of the latter.

Q.E.D. (tertio loco)

Saccheri's general conclusion, therefore, was:

> It is established that the join *CD* will be equal to, or less than, or greater than this *AB*, according as the angles at the same *CD* are right, or obtuse, or acute.

Proposition IV is the converse of Proposition III. It is stated as follows:

> But inversely (Figure 3) the angles at the join *CD* will be right, or obtuse, or acute, according as the straight *CD* is equal, or less, or greater than the opposite *AB*.

> Proof: For if the straight *CD* is equal to the opposite *AB*, and nevertheless the angles at it are either obtuse or acute; now these such angles prove it (Prop. III) not equal, but less, or greater than the opposite *AB*; which is absurd against the hypothesis.

> The same uniformly avails in regard to the remaining cases. It holds therefore that the angles at the join *CD* are either right, or obtuse, or acute, according as the straight *CD* is equal to, or less than, or greater than the opposite *AB*.

Q.E.D.

Saccheri called the angles at *C* and *D* the *summit* angles of the quadrilateral, and, as we see from the first four propositions, noted the three following possibilities:

1. The summit angles are right (the right-angle hypothesis).

2. The summit angles are obtuse (the obtuse-angle hypothesis).

3. The summit angles are acute (the acute-angle hypothesis).

In Propositions V, VI, and VII, Saccheri proved that if one of these hypotheses is true for one of his quadrilaterals, it is true for every such quadrilateral. Using the Postulate of Archimedes, which includes implicitly the infinitude of the straight line, Saccheri showed, in Proposition VI, that the Fifth Postulate is a consequence of the right-angle hypothesis, and finally, in Proposition XIV, that the obtuse-angle hypothesis is self-contradictory.

It now remained for Saccheri to dispose of the acute-angle hypothesis, which he explored in the hope of finding a contradiction. Obtaining many results that seemed strange because they differed to a great degree from those that had been established by use of the Fifth Postulate, he never succeeded in finding the desired contradiction. Since he found it impossible to cast out the acute-angle hypothesis on purely logical grounds and so "prove" the Fifth Postulate, Saccheri concluded in Proposition XXXIII of his famous book, *Euclides ab omni naevo vindicatus (Euclid Freed of Every Flaw)*, that the "hypothesis of the acute angle is absolutely false because repugnant to the nature of the straight line." For this proof he trusted to intuition and to faith in the validity of the Fifth Postulate rather than to logic, and relied on five Lemmas spread over sixteen pages of his masterpiece, "in which are contained five fundamental axioms relating to the straight line and circle, with their correlative postulates."

It is highly improbable that Saccheri was satisfied with his investigation, for he was an able logician. Did he really expect to find a flaw in Euclid's fundamental postulates, or did he have a preconceived idea that no defect was to be found in his idol? Why the particular title, *Euclid Freed of Every Flaw?* Vailati hints that Saccheri was already convinced before beginning his proofs that he would find no contradiction, and that the sole purpose of his work was to "free Euclid." He points out that Saccheri already, in his youth, had arrived at the idea that the characteristic property of the most fundamental propositions in every demonstrative science is precisely their indemonstratibility, except by assuming as hypothesis the falsity of the very proposition to be proved, and in showing how also, when taking this hypothesis as point of departure, one arrives at the conclusion that the proposition in question is true.

Vailati goes on to say, "It was in hopes of reaching in this way a proof of the Parallel Postulate, namely, deducing it from the very hypothesis of its falsity, that Saccheri pushed on in the investigation of the consequences flowing from the other two alternative hypotheses, to which the negation of the Parallel Postulate gave rise, attaining thus results fitting to carry on in their sequence to a discovery far more important than what he had in mind to reach, namely to the discovery of a wholly new geometry of which the old is only a simple particular case."

"In this regard," continues Vailati, "his position is not unlike that of his fellow countryman, Columbus, who precisely in the hopes of reaching by a new way regions already known, was led to the discovery of a new continent."

Dr. Withers, in his book, *Euclid's Parallel Postulate*, published in 1908, attributes Saccheri's conclusion not to a logical but to a psychological difficulty. He states: "We have seen that the assumption (acuti) worried Saccheri very profoundly in his heroic efforts to 'vindicate Euclid.' It was not a logical but a psychological or experiential difficulty which caused Saccheri to reject the logical conclusions to which his own labors clearly and inevitably pointed; and it was certainly the same sort of difficulty which caused the immediate rejection, by himself and subsequent mathematicians, of the assumption (obtusi)."

Although Saccheri's work, *Euclid Freed of Every Flaw*, failed in its aim, it is of great importance. In it the most determined effort had been made on behalf of the Fifth Postulate, and the fact that Saccheri did not succeed in discovering any contradictions among the consequences of the *hypothesis of the acute angle* could hardly help but suggest the question as to whether a consistent logical geometrical system could be built upon the hypothesis, and thus the Euclidean Postulate be impossible of demonstration.

The publication of Saccheri's *Euclides ab omni naevo vindicatus*, in 1733, attracted some attention. Mention is made of it in two early histories of

mathematics, that of J. C. Heilbronner (Leipzig, 1742), and that of Montucla (Paris, 1758). It was also noted by G. S. Klügel, in his dissertation (Göttingen, 1763). But then it seems to have been forgotten until it was accidentally rediscovered by the Jesuit Father Angelo Manganotti in 1889. It is not generally known whether Father Manganotti had previous knowledge of Saccheri and his work or if his first acquaintance with him was through Beltrami, who in 1889 named Saccheri as "the Italian precursor of the Hungarian Bolyai and the Russian Lobachevski."

Following the rediscovery of Saccheri's work, it was translated from the Latin into English by Dr. George Bruce Halsted of the University of Texas, into German by Engel and Stäckel, and into Italian by Boccardini. But with all of this, there were hundreds of mathematicians who knew nothing of this publication. It might be noted here that even Sir Thomas Heath, in his famous three-volume *Euclid's Elements* (Cambridge, 1908), referred to Saccheri's work as a Latin translation of Euclid.

What influence did Saccheri exert on the geometers of the eighteenth century? It is difficult to say. However, Roberto Bonola, in his *Non-Euclidean Geometry* (1906), shows that Saccheri had a very decided influence on a few of the leading mathematicians of that period. The first was Johann Heinrich Lambert (1728–1777), a Swiss mathematician who quotes, in his *Theorie der Parallellinien*, a dissertation of G. S. Klügel (1739–1812) in which the work of the Italian geometer Saccheri is carefully analyzed. Part of Lambert's investigation clearly resembles that of Saccheri. His fundamental figure is a quadrilateral with three right angles. The *three* hypotheses are made concerning the nature of the third angle, and in the treatment the author does not depart from the Saccheri method.

Adrien Marie Legendre (1752–1853), a French mathematician, also displayed knowledge of Saccheri's work. In his *Eléments de géométrie*, his investigations of the theory of parallels were much like Saccheri's, and the results he obtained were to a large extent the same. He chose, however, to place emphasis upon the angle-sum of a triangle and proposed three hypotheses in which the sum of the angles is, in turn, equal to, greater than, and less than two right angles, hoping to be able to reject the last two. Unconsciously assuming the straight line infinite, he was able to eliminate the geometry based on the second hypothesis by proving the following theorem: The sum of the angles of a triangle cannot be greater than two right angles.

Even though Legendre added nothing new to the materials and results obtained by Saccheri and Lambert, yet the simple, straightforward style of his proofs brought him a large following and helped to create an interest in these ideas just at a time when geometers were on the threshold of a great discovery. Some of his proofs, on account of their elegance, are of permanent value. In spite of his many attempts, he was never able to dispose of the third hypothesis, which, as Gauss remarked, was "the reef on which all the wrecks occurred."

So far, the term "non-Euclidean geometry" had not been used, and it remained for Gauss to bring that expression into our mathematical language. Carl Friedrich Gauss (1777–1855), like Saccheri and Lambert before him, had attempted to prove the truth of Euclid's Fifth Postulate by assuming its falsity. Contrary to popular opinion, he did not recognize the existence of a logically sound non-Euclidean geometry by intuition or a flash of genius, but rather he spent many laborious years before he had overcome the inherited prejudice against it. How much influence was exerted on Gauss by Saccheri is best expressed by Segre in his "Congetture intorno alla influenzadi Saccheri sulla formazione della geometria non euclidea" (1903). He remarks that both Gauss and Wolfgang Bolyai, while students at Göttingen, the former from 1795 to 1798 and the latter from 1796 to 1799, were interested in the theory of parallels. It is therefore possible that, through their professors Kästner and Seyffer, who were both deeply versed in this subject, they had become familiar with both Saccheri's *Euclides vindicatus* and Lambert's *Theorie*

der Parallellinien. There is no confirmation of this, however; it is only a conjecture.

In one of his letters to his friend Schumacher, Gauss mentioned a "certain Schweikart." The one referred to was Ferdinand Karl Schweikart (1780–1859), who from 1796 to 1798 was a student of law at Marburg. As he was keenly interested in mathematics he took advantage of the opportunity while at the university to listen to various lectures on the subject, and particularly to those of J. K. F. Hauff, who was somewhat of an authority on the theory of parallels. Schweikart's interest in this theory developed to such an extent that in 1807 there appeared his only published work of a mathematical nature, *Die Theorie der Parallellinien nebst dem Vorschlage ihrer Verbannung aus der Geometrie.*

In this book Schweikart mentioned both Saccheri and Lambert, but the contents were in no way novel and the style was on quite conventional lines. Doubtless his acquaintance with the work of Saccheri and Lambert affected the character of his later investigations. Eleven years later, having discovered a new order of ideas, Schweikart developed a geometry independent of Euclid's hypothesis. So, in 1818, he sent the following memorandum to Gauss for the latter's opinion: ". . . There are two kinds of geometry—a geometry in the strict sense, the Euclidean; and an Astral Geometry. . . . ." The "astral geometry," as discovered by Schweikart, as well as Gauss's "non-Euclidean geometry," corresponds exactly to Saccheri's system for the hypothesis of the acute angle. And since Schweikart in his *Theorie* of 1807 mentioned the work of both Saccheri and Lambert, we can easily assume the influence exerted upon him by these earlier works.

It should not surprise us that the discovery of non-Euclidean geometry was not made by one person, but independently by several in different parts of the world. We know from the history of mathematics that this has happened on more than one occasion, and undoubtedly will happen again. Slowly but surely, the efforts of such men as Saccheri, Lambert, and others, in the investigation of the Fifth Postulate directed the speculations of geometers to the point where the discovery was imminent.

Gauss, whose courage did not match his genius, failed to complete his discoveries. He did, however, keep in touch with the Hungarian mathematician, Wolfgang Bolyai (1775–1856), who had been his fellow-student in Göttingen. The latter's son, János, in 1823 developed a new system of geometry on the "acute angle hypothesis," in which those theorems of Euclid that are independent of the Fifth Postulate still operated, but in which the others were replaced by such amazing conclusions as this: In a plane, instead of one line, two lines can be drawn parallel to a given line, and through this point, an infinite number of lines may be drawn lying in the angle between the first two and having the property that they will not intersect the given line.

Amazing as this is in contrast to Euclidean geometry, it is equally logical. The younger Bolyai's discovery was published in 1832 as a supplement to a work by his father, to the great delight of Gauss, who stated that this publication relieved him of all responsibility to complete his own work. But, as we know today, Bolyai's work was almost completely ignored, and hence, soon forgotten.

Meanwhile, in Russia, a little-known mathematician named Nikolai Lobachevski (1793–1856) was lecturing on the "acute angle geometry" at the University of Kazan. When the German version of his work was published in 1840, Gauss was again enthusiastic, but Bolyai declared that Lobachevski had been strongly influenced by his own publication of 1832. It is now well known that Lobachevski's work had been presented in Russian at least eleven years prior to the German publication. If any influence was exerted on Lobachevski, it was probably through one of his teachers, Johann Martin Christian Bartels (1769–1836), a friend of Gauss who went to Kazan in 1807 and became professor of mathematics at that university.

Some notes of Lobachevski, that were written during the period 1815–1817, show his attempts at the proof of the Fifth Postulate, but his investigations seemed to resemble those of Legendre. Whether this was a result of an influence passed on to him from the works of Gauss or even Saccheri, or whether the resemblance is purely coincidental, is difficult to assume. One thing, however, is certain: his discoveries were not favorably received, and the story of his treatment at the hands of the government is given at the beginning of this paper.

Neither Bolyai nor Lobachevski gave any consideration to the "obtuse angle hypothesis." But Georg Friedrich Bernhard Riemann (1826–1866), a student of Gauss in Göttingen, began with a study of the postulate that a straight line is infinitely long, the assumption that had convinced Saccheri that the "obtuse angle hypothesis" is untenable. Discarding this postulate, Riemann developed a geometry in which all lines are of finite length, any pair of lines intersect if they lie in the same plane, and the sum of the angles of a triangle is greater than two right angles.

Reimann, in his work, called attention to the true nature and significance of geometry, and did much to free mathematics from the handicap of tradition. In fact, his work dealt almost entirely with generalities, and was suggestive in nature, so as to leave opportunities for detailed investigations to be carried on later by his successors in the field of non-Euclidean geometry.

Earlier in this paper it was stated that the term "non-Euclidean Geometry" was first used by Gauss to denote that geometry which arises on replacing Euclid's Fifth Postulate by its negation, and keeping unaltered all the remaining postulates that Euclid either explicitly formulated or which enter implicitly into his development of geometry. In this sense, there is only one non-Euclidean geometry, namely, Saccheri's geometry of the "acute-angle hypothesis." But Felix Klein, in 1871, gave a different nomenclature to the existing geometries. That which had been investigated by Saccheri, Gauss, Bolyai, and Lobachevski, he called "hyperbolic geometry"; that of Riemann, he called "elliptic geometry"; while to Euclid's he gave the name of "parabolic geometry." These names were suggested by the fact that a straight line contains two infinitely distant points under the *hypothesis of the acute angle*, none under the *hypothesis of the obtuse angle*, and only one under the *hypothesis of the right angle*. The characteristic postulate of Euclidean geometry states that: "Through a given point, one and only one line can be drawn which is parallel to a given line." On the other hand, the distinctive feature of hyperbolic plane geometry is the assumption that an infinite number of parallels can be drawn to a line through a point. An investigation of the supposition that no line can be drawn through a point parallel to a given line brings us to the hypothesis of the obtuse angle of Saccheri, which geometry he ruled out, because he assumed that straight lines are infinite. It was Riemann who first pointed out the importance of distinguishing between the ideas of *boundlessness* and *infinitude* in connection with the concepts of space. Hence, the transition from Euclidean to elliptic geometry is not a very simple task, and elliptic geometry's characteristic postulate, "Two straight lines always intersect with each other," is not only incompatible with the Euclidean postulate which it replaces, but also with others.

So, at last, slowly but surely, the obstinate puzzle of the Fifth Postulate was being solved. One wonders why this should have taken so long and why the chain reaction set off by Saccheri had not fissioned more violently. But one must remember that during this period the philosophy of Kant (1724–1804), which treated space not as empirical, but as intuitive, dominated the situation. Space was regarded as something existing in the mind, not as a concept resulting from external experience. It required courage at that time to recognize that geometry was an experimental science once it was applied to physical space, and that its postulates and their consequences need only be accepted if convenient, and if they agree reasonably well with experimental data. It was the domination of the Kantian philosophy which dampened Gauss's courage and which influenced the public against the discoveries of Bolyai and Lobachevski.

Gradually a change of viewpoint arrived and the new discovery led to the complete destruction of the Kantian space concept and the revelation of the true distinction between concept and experience, as well as their interrelation. Now that the term "non-Euclidean geometry" has become almost a household expression, many of us may find it difficult to reconcile the attitude of mathematicians of greater and lesser stature over the period of transition. Again, this attitude should not be too much of a shock to us. The history of scientific discovery teaches that every radical change in its separate departments does not suddenly alter the convictions and the presuppositions upon which investigators and teachers have for a considerable time based the presentation of their subjects. Were Saccheri to return today to this new world, he would probably be the first to acknowledge that one step would have brought him to the full realization of the discovery that stared him in the face as he made his heroic attempt to free his idol, Euclid, from every blemish. When he died on October 25, 1733, at the age of sixty-six, it is doubtful if he had any intimation of the fire he had ignited, and of the future generations who would hail him as "the forerunner of non-Euclidean geometry."

BIBLIOGRAPHY

Bell, Eric T. *Mathematics, Queen and Servant of Science.* New York: McGraw-Hill Book Co., Inc., 1951.

———. *Men of Mathematics.* New York: Simon and Schuster, Inc., 1937.

*———. *The Development of Mathematics.* New York: McGraw-Hill Book Co., Inc., 1945.

———. *The Magic of Numbers.* New York: McGraw-Hill Book Co., Inc., 1946.

*Blumenthal, Leonard M. *A Modern View of Geometry.* San Francisco: W. H. Freeman & Co., Publishers, 1961.

*Available as a Dover reprint.

Bonola, Roberto. *Non-Euclidean Geometry* (1906). Translated by H. S. Carslaw, Sydney, 1911. New York: Dover Publications, 1955.

Cajori, Florian. *A History of Mathematics.* New York: The Macmillan Company, 1938.

Eves, Howard. *An Introduction to the History of Mathematics,* revised edition. New York: Holt, Rinehart & Winston, Inc., 1964.

Hofmann, Joseph E. *Classical Mathematics.* New York: Philosophical Library, Inc., 1959.

Kattsoff, Louis O. "The Saccheri Quadrilateral," *The Mathematics Teacher,* LV (December, 1962): 630-36.

Kline, Morris. *Mathematics in Western Culture.* New York: Oxford University Press, Inc., 1953.

Kramer, Edna E. *The Main Stream of Mathematics.* New York: Oxford University Press, Inc., 1953.

Miller, George A. *Historical Introduction to Mathematical Literature.* New York: The Macmillan Company, 1916.

*Newman, James R. (ed.) *The World of Mathematics,* Vol. III. New York: Simon and Schuster, Inc., 1956.

Saccheri, Girolamo G. *Euclides ab omni naevo vindicatus* (1733). Translated by George B. Halsted. Chicago: Open Court Publishing Co., 1920.

Sanford, Vera. *A Short History of Mathematics.* Boston: Houghton Mifflin Company, 1930.

*Smith, David E., *History of Mathematics,* Vol. II. Boston: Ginn & Company, 1925.

*———. *A Source Book in Mathematics.* New York: McGraw-Hill Book Co., Inc., 1929.

Wolfe, Harold E. *Introduction to Non-Euclidean Geometry.* New York: Holt, Rinehart & Winston, Inc., 1945.

*Available as a Dover reprint.

Synopsis of Major Geometries

Founder	Popular Name	Possible Model	Number of Lines Parallel to a Given Line	Sum of the Interior Angles of a Triangle
Euclid (ca. 300 B.C.)	Euclidean	A plane (a)	only one	180°
N. Lobachevsky (1830) J. Bolyai (1832)	Hyperbolic	Poincaré circle (b)	many	less that 180°
G. Riemann	Elliptic	Sphere (c)	none	greater than 180°

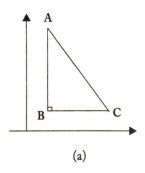

(a)

$\angle A + \angle B + \angle C = 180°$

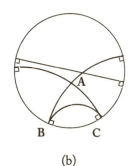

(b)

$\angle A + \angle B + \angle C < 180°$

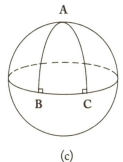

(c)

$\angle A + \angle B + \angle C > 180°$

Poincaré Circle

Given a circle, lines drawn within this circle have the following properties: any line passing through the center of the circle will continue to the circumference as a "straight" line; lines not through the center will be circular arcs orthogonal to the circumference.

The Evolution of Group Theory: A Brief Survey

ISRAEL KLEINER

THIS ARTICLE gives a brief sketch of the evolution of group theory. It derives from a firm conviction that the history of mathematics can be a useful and important integrating component in the teaching of mathematics. This is not the place to elaborate on the role of history in teaching, other than perhaps to give one relevant quotation:

> Although the study of the history of mathematics has an intrinsic appeal of its own, its chief raison d'être is surely the illumination of mathematics itself. For example the gradual unfolding of the integral concept from the volume computations of Archimedes to the intuitive integrals of Newton and Leibniz and finally to the definitions of Cauchy, Riemann and Lebesgue—cannot fail to promote a more mature appreciation of modern theories of integration.
>
> —C. H. Edwards

The presentation in one article of the evolution of so vast a subject as group theory necessitated severe selectivity and brevity. It also required omission of the broader contexts in which group theory evolved, such as wider currents in abstract algebra, and in mathematics as a whole. (We will note *some* of these interconnections shortly.) We trust that enough of the essence and main lines of development in the evolution of group theory have been retained to provide a useful beginning from which the reader can branch out in various directions. For this the list of references will prove useful.

The reader will find in this article an outline of the origins of the main concepts, results, and theories discussed in a beginning course on group theory. These include, for example, the concepts of (abstract) group, normal subgroup, quotient group, simple group, free group, isomorphism, homomorphism, automorphism, composition series, direct product; the theorems of J. L. Lagrange, A.-L. Cauchy, A. Cayley, C. Jordan-O. Hölder; the theories of permutation groups and of abelian groups. At the same time we have tried to balance the technical aspects with background information and interpretation.

Our survey of the evolution of group theory will be given in several stages, as follows:

1. Sources of group theory.
2. Development of "specialized" theories of groups.
3. Emergence of abstraction in group theory.
4. Consolidation of the abstract group *concept;* dawn of abstract group *theory.*
5. Divergence of developments in group theory.

Reprinted from *Mathematics Magazine* 59 (Oct., 1986): 195–215; with permission of the Mathematical Association of America and the author.

Before dealing with each stage in turn, we wish to mention the context within mathematics as a whole, and within algebra in particular, in which group theory developed. Although our "story" concerning the evolution of group theory begins in 1770 and extends to the 20th century, the major developments occurred in the 19th century. Some of the general mathematical features of that century which had a bearing on the evolution of group theory are: (a) an increased concern for rigor; (b) the emergence of abstraction; (c) the rebirth of the axiomatic method; (d) the view of mathematics as a human activity, possible without reference to, or motivation from, physical situations. Each of these items deserves extensive elaboration, but this would go beyond the objectives (and size) of this paper.

Up to about the end of the 18th century, algebra consisted (in large part) of the study of solutions of polynomial equations. In the 20th century, algebra became a study of abstract, axiomatic systems. The transition from the so-called classical algebra of polynomial equations to the so-called modern algebra of axiomatic systems occurred in the 19th century. In addition to group theory, there emerged the structures of commutative rings, fields, noncommutative rings, and vector spaces. These developed alongside, and sometimes in conjunction with, group theory. Thus Galois theory involved both groups and fields; algebraic number theory contained elements of group theory in addition to commutative ring theory and field theory; group representation theory was a mix of group theory, noncommutative algebra, and linear algebra.

Sources of Group Theory

There are four major sources in the evolution of group theory. They are (with the names of the originators and dates of origin):

(a) Classical algebra (J. L. Lagrange, 1770)
(b) Number theory (C. F. Gauss, 1801)
(c) Geometry (F. Klein, 1874)

(d) Analysis (S. Lie, 1874; H. Poincaré and F. Klein, 1876)

We deal with each in turn.

(a) *Classical Algebra* (J. L. Lagrange, 1770)

The major problems in algebra at the time (1770) that Lagrange wrote his fundamental memoir "Réflexions sur la résolution algébrique des équations" concerned polynomial equations. There were "theoretical" questions dealing with the existence and nature of the roots (e.g., Does every equation have a root? How many roots are there? Are they real, complex, positive, negative?), and "practical" questions dealing with methods for finding the roots. In the latter instance there were exact methods and approximate methods. In what follows we mention exact methods.

The Babylonians knew how to solve quadratic equations (essentially by the method of completing the square) around 1600 B.C. Algebraic methods for solving the cubic and the quartic were given around 1540. One of the major problems for the next two centuries was the algebraic solution of the quintic. This is the task Lagrange set for himself in his paper of 1770.

In his paper Lagrange first analyzes the various known methods (devised by F. Viète, R. Descartes, L. Euler, and E. Bézout) for solving cubic and quartic equations. He shows that the common feature of these methods is the reduction of such equations to auxiliary equations—the so-called resolvent equations. The latter are one degree lower than the original equations. Next Lagrange attempts a similar analysis of polynomial equations of arbitrary degree n. With each such equation he associates a "resolvent equation" as follows: let $f(x)$ be the original equation, with roots $x_1, x_2, \ldots, x_n$. Pick a rational function $R(x_1, x_2, \ldots, x_n)$ of the roots and coefficients of $f(x)$. (Lagrange describes methods for doing so.) Consider the different values which $R(x_1, x_2, \ldots, x_n)$ assumes under all the

Evariste Galois

$n!$ permutations of the roots $x_1, x_2,\ldots, x_n$ of $f(x)$. If these are denoted by $y_1, y_2,\ldots, y_k$, then the resolvent equation is given by $g(x) = (x - y_1).(x - y_2) \ldots (x - y_k)$. (Lagrange shows that k divides $n!$—the source of what we call Lagrange's theorem in group theory.) For example, if $f(x)$ is a quartic with roots x_1, x_2, x_3, x_4, then $R(x_1, x_2, x_3, x_4)$ may be taken to be $x_1, x_2, + x_3, x_4$, and this function assumes three distinct values under the 24 permutations of x_1, x_2, x_3, x_4. Thus the resolvent equation of a quartic is a cubic. However, in carrying over this analysis to the quintic, he finds that the resolvent equation is of degree six!

Although Lagrange did not succeed in resolving the problem of the algebraic solvability of the quintic, his work was a milestone. It was the first time that an association was made between the solutions of a polynomial equation and the permutations of its roots. In fact, the study of the permutations of the roots of an equation was a cornerstone of Lagrange's general theory of algebraic equations. This, he speculated, formed "the true principles for the solution of equations.(He was, of course, vindicated in this by E. Galois.) Although Lagrange speaks of permutations without considering a "calculus"of permutations (e.g., there is no consideration of their composition or closure), it can be said that the germ of the group concept (as a group of permutations) is present in his work.

(b) *Number Theory* (C. F. Gauss, 1801)

In the *Disquisitiones Arithmeticae* of 1801 Gauss summarized and unified much of the number theory that preceded him. The work also suggested new directions which kept mathematicians occupied for the entire century. As for its impact on group theory, the *Disquisitiones* may be said to have initiated the theory of finite abelian groups. In fact, Gauss established many of the significant properties of these groups without using any of the terminology of group theory. The groups appear in four different guises: the additive group of integers modulo m, the multiplicative group of integers relatively prime to m, modulo m, the group of equivalence classes of binary quadratic forms, and the group of nth roots; of unity. And though these examples appear in number-theoretic contexts, it is as abelian groups that Gauss treats them, using what are clear prototypes of modern algebraic proofs.

For example, considering the nonzero integers modulo p (p a prime), Gauss shows that they are all powers of a single element; i.e., that the group Z_p^* of such integers is cyclic. Moreover, he determines the number of generators of this group (he shows that it is equal to $\phi(p-1)$, where ϕ is Euler's ϕ-function). Given any element of Z_p^* he defines the order of the element (without using the terminology) and shows that the order of an element is a divisor of $p-1$. He then uses this result to prove P. Fermat's "little theorem," namely, that $a^{p-1} \equiv 1 \bmod p$ if p does not divide a, thus employing group-theoretic ideas to prove number-theoretic results. Next he shows that if t is a positive integer which divides $p-1$, then there exists an element in Z_p^* whose order is t— essentially the converse of Lagrange's theorem for cyclic groups.

Concerning the nth roots of 1 (which he considers in connection with the cyclotomic equation), he shows that they too form a cyclic group. In connection with this group he raises and answers many of the same questions he raised and answered in the case of Z_p^*.

The problem of representing integers by binary quadratic forms goes back to Fermat in the early 17th century. (Recall his theorem that every prime of the form $4n + 1$ can be represented as a sum of two squares $x^2 + y^2$.) Gauss devotes a large part of the Disquisitiones to an exhaustive study of binary quadratic forms and the representation of integers by such forms. (A **binary quadratic form** is an expression of the form $ax^2 + bxy + cy^2$, with a, b, c integers.) He defines a composition on such forms, and remarks that if K and K^1 are two such forms one may denote their composition by $K + K^1$. He then shows that this composition is associative and commutative, that there exists an identity, and that each form has an inverse, thus verifying all the properties of an abelian group.

Despite these remarkable insights one should not infer that Gauss had the concept of an abstract group, or even of a finite abelian group. Although the arguments in the *Disquisitiones* are quite general, each of the various types of "groups" he considers is dealt with separately— there is no unifying group-theoretic method which he applies to all cases.

(c) *Geometry* (F. Klein, 1872)

We are referring here to Klein's famous and influential lecture entitled "A Comparative Review of Recent Researches in Geometry," which he delivered in 1872 on the occasion of his admission to the faculty of the University of Erlangen. The aim of this so-called Erlangen Program was the classification of geometry as the study of invariants under various groups of transformations. Here there appear groups such as the projective group, the group of rigid motions, the group of similarities, the hyperbolic group, the elliptic groups, as well as the geometries associated with them. (The affine group was not mentioned by Klein.) Now for some background leading to Klein's Erlangen Program.

The 19th century witnessed an explosive growth in geometry, both in scope and in depth. New geometries emerged: projective geometry, noneuclidean geometries, differential geometry, algebraic geometry, η-dimensional geometry, and Grassmann's geometry of extension. Various geometric methods competed for supremacy: the synthetic versus the analytic, the metric versus the projective. At mid-century, a major problem had arisen, namely, the classification of the relations and inner connections among the different geometries and geometric methods. This gave rise to the study of "geometric relations," focusing on the study of properties of figures invariant under transformations. Soon the focus shifted to a study of the transformations themselves. Thus the study of the geometric relations of figures became the study of the associated transformations. Various types of transformations (e.g., collineations, circular transformations, inversive transformations, affinities) became the objects of specialized studies. Subsequently, the logical connections among transformations were investigated, and this led to the problem of classifying transformations and eventually to Klein's group-theoretic synthesis of geometry.

Klein's use of groups in geometry was the final stage in bringing order to geometry. An intermediate stage was the founding of the first major theory of classification in geometry, beginning in the 1850's, the Cayley-Sylvester Invariant Theory. Here the objective was to study invariants of "forms" under transformations of their variables. This theory of classification, the precursor of Klein's Erlangen Program, can be said to be *implicitly* group-theoretic. Klein's use of groups in geometry was, of course, explicit. In the next section (2−(c)) we will note the significance of

Felix Klein

Klein's Erlangen Program (and his other works) for the evolution of group theory. Since the Program originated a hundred years after Lagrange's work and eighty years after Gauss' work, its importance for group theory can best be appreciated after a discussion of the evolution of group theory beginning with the works of Lagrange and Gauss and ending with the period around 1870.

(d) *Analysis* (S. Lie, 1874; H. Poincaré and F. Klein, 1876)

In 1874 Lie introduced his general theory of (continuous) transformation groups—essentially what we call Lie groups today. Such a group is represented by the transformations

$$x'_i = f_i(x_1, x_2, ..., x_n, a_1, a_2, ..., a_n), \qquad i = 1, 2, ..., n,$$

where the f are analytic functions in the x_i, and a_i (the a_i are parameters, with both x_i and a_i real or complex). For example, the transformations given by

$$x' = \frac{ax + b}{cx + d},$$

where a, b, c, d, are real numbers and $ad - bc \neq 0$,

define a continuous transformation group.

Lie thought of himself as the successor of N. H. Abel and Galois, doing for differential equations what they had done for algebraic equations. His work was inspired by the observation that almost all the differential equations which had been integrated by the older methods remain invariant under continuous groups that can be easily constructed. He was then led to consider, in general, differential equations that remain invariant under a given continuous group and to investigate the possible simplifications in these equations which result from the known properties of the given group (cf. Galois theory). Although Lie did

Sophus Lie

not succeed in the actual formulation of a "Galois theory of differential equations," his work was fundamental in the subsequent formulation of such a theory by E. Picard (1883/1887) and E. Vessiot (1892).

Poincaré and Klein began their work on "automorphic functions" and the groups associated with them around 1876. Automorphic functions (which are generalizations of the circular, hyperbolic, elliptic, and other functions of elementary analysis) are functions of a complex variable z, analytic in some domain D, which are invariant under the group of transformations

$$z' = \frac{az+b}{cz+d}, \quad \begin{array}{l} (a, b, c, d \text{ real or complex and} \\ ad - bc \neq 0) \end{array}$$

or under some subgroup of this group. Moreover, the group in question must be "discontinuous" (i.e., any compact domain contains only finitely many transforms of any point). Examples of such groups are the modular group (in which a, b, c, d are integers and $ad - bc = 1$), which is associated with the elliptic modular functions, and Fuchsian groups (in which a, b, c, d are real and $ad - bc = 1$) associated with the Fuchsian automorphic functions. As in the case of Klein's Erlangen Program, we will explore the consequences of these works for group theory in section 2–(c).

Development of "Specialized" Theories of Groups

In §1 we outlined four major sources in the evolution of group theory. The first source—classical algebra—led to the theory of permutation groups; the second source—number theory—led to the theory of abelian groups; the third and fourth sources—geometry and analysis—led to the theory of transformation groups. We will now outline some developments within these specialized theories.

(a) Permutation Groups

As noted earlier, Lagrange's work of 1770 initiated the study of permutations in connection with the study of the solution of equations. It was probably the first clear instance of implicit group-theoretic thinking in mathematics. It led directly to the works of P. Ruffini, Abel, and Galois during the first third of the 19th century, and to the concept of a permutation group.

Ruffini and Abel proved the unsolvability of the quintic by building upon the ideas of Lagrange concerning resolvents. Lagrange showed that a necessary condition for the solvability of the general polynomial, equation of degree n is the existence of a resolvent of degree less than n. Ruffini and Abel showed that such resolvents do not exist for $n > 4$. In the process they developed a considerable amount of permutation theory. It was Galois, however, who made the fundamental conceptual advances, and who is considered by many as the founder of (permutation) group theory.

Galois' aim went well beyond finding a method for solvability of equations. He was concerned with gaining insight into general principles, dissatisfied as he was with the methods of his predecessors: "From the beginning of this century," he wrote, "computational procedures have become so complicated that any progress by those means has become impossible."

Galois recognized the separation between "Galois theory" (i.e., the correspondence between fields and groups) and its application to the solution of equations, for he wrote that he was presenting "the general principles and just one application" of the theory. "Many of the early commentators on Galois theory failed to recognize this distinction, and this led to an emphasis on applications at the expense of the theory" (Kiernan).

Galois was the first to use the term "group" in a technical sense—to him it signified a collection of permutations closed under multiplication: "if one has in the same group the substitutions S and T one is certain to have the substitution ST." He recognized that the most important properties of an algebraic equation were reflected in certain properties of a group uniquely associated with the equation— "the group of the equation." To describe these properties he invented the fundamental notion of normal subgroup and used it to great effect. While the issue of resolvent equations preoccupied Lagrange, Ruffini, and Abel, Galois' basic idea was to bypass them, for the construction of a resolvent required great skill and was not based on a clear methodology. Galois noted instead that the existence of a resolvent was equivalent to the existence of a normal subgroup of prime index in the group of the equation. This insight shifted consideration from the resolvent equation to the group of the equation and its subgroups.

Galois defines the group of an equation as follows:

> Let an equation be given, whose m roots are a, b, c, ... There will always be a group of permutations of the letters a, b, c, ... which has the following property: 1) that every function of the roots, invariant under the substitutions of that group, is rationally known [i.e., is a rational function of the coefficients and any adjoined quantities], 2) conversely, that every function of the roots, which can be expressed rationally, is invariant under these substitutions.

The definition says essentially that the group of the equation consists of those permutations of the roots of the equation which leave invariant all relations among the roots over the field of coefficients of the equation—basically the definition we would give today. Of course the definition does not guarantee the existence of such a group, and so Galois proceeds to demonstrate it. Galois next investigates how the group changes when new elements are adjoined to the "ground field" F. His treatment is amazingly close to the standard treatment of this matter in a modern algebra text.

Galois' work was slow in being understood and assimilated. In fact, while it was done around 1830, it was published posthumously in 1846, by J. Liouville. Beyond his technical accomplishments, Galois "challenged the development of mathematics in two ways. He discovered, but left unproved, theorems which called for proofs based on new, sophisticated concepts and calculations. Also, the task of filling the gaps in his work necessitated a fundamental clarification of his methods and their group theoretical essence" (Wussing).

The other major contributor to permutation theory in the first half of the 19th century was Cauchy. In several major papers in 1815 and 1844 Cauchy inaugurated the theory of permutation groups as an autonomous subject. (Before Cauchy, permutations were not an object of independent study but rather a useful device for the investigation of solutions of polynomial equations.) Although Cauchy was well aware of the work of Lagrange and Ruffini (Galois' work was not yet published at the time), Wussing suggests that Cauchy "was definitely not inspired directly by the contemporary group-theoretic formulation of the solution of algebraic equations."

In these works Cauchy gives the first systematic development of the subject of permutation groups. In the 1815 papers Cauchy uses no special name for sets of permutations closed under multiplication. However, he recognizes their importance and gives a name to the number of elements in such a closed set, calling it "diviseur indicatif" In the 1844 paper he defines the concept of a group of permutations generated by certain elements.

> Given one or more substitutions involving some or all of the elements x, y, z,... I call the products of these substitutions, by themselves or by any other, in any order, *derived* substitutions. The given substitutions, together with the derived ones, form what I call a *system of conjugate substitutions*.

In these works, which were very influential, Cauchy makes several lasting additions to the terminology, notation, and results of permutation theory. For example, he introduces the permutation notation $\begin{pmatrix} x & y & z \\ x & y & z \end{pmatrix}$ in use today, as well as the cyclic notation for permutations; defines the product of permutations, the degree of a permutation, cyclic permutation, transposition; recognizes the identity permutation as a permutation; discusses what we would call today the direct product of two groups; and deals with the alternating groups extensively. Here is a sample of some of the results he proves.

(i) Every even permutation is a product of 3-cycles.

(ii) If p (prime) is a divisor of the order of a group, then there exists a subgroup of order p. (This is known today as "Cauchy's theorem", though it was stated without proof by Galois.)

(iii) Determined all subgroups of S_3, S_4, S_5, S_6 (making an error in S_6.)

(iv) All permutations which commute with a given one form a group (the centralizer of an element).

It should be noted that all these results were given and proved in the context of permutation groups.

The crowning achievement of these two lines of development—a symphony on the grand themes of Galois and Cauchy—was Jordan's important and influential *Traité des substitutions et des équations algébriques* of 1870. Although the author states in the preface that "the aim of the work is to develop Galois' method and to make it a proper field of study, by showing with what facility it can solve all principal problems of the theory of equations," it is in fact group theory per se—not as an offshoot of the theory of solvability of equations—which forms the central object of study.

Camille Jordan

The striving for a mathematical synthesis based on key ideas is a striking characteristic of Jordan's work as well as that of a number of other mathematicians of the period (e.g., F. Klein). The concept of a (permutation) group seemed to Jordan to provide such a key idea. His approach enabled him to give a unified presentation of results due to Galois, Cauchy, and others. His application of the group concept to the theory of equations, algebraic geometry, transcendental functions, and theoretical mechanics was also part of the unifying and synthesizing theme. "In his book Jordan wandered through all of algebraic geometry, number theory, and function theory in search of interesting permutation groups" (Klein)." In fact, the aim was a survey of all of mathematics by areas in which the theory of permutation groups had been applied or seemed likely to be applicable. "The work represents . . . a review of the whole of contemporary mathematics from the standpoint of the occurrence of group-theoretic thinking in permutation-theoretic form" (Wussing).

The *Traité* embodied the substance of most of Jordan's publications on groups up to that time (he wrote over 30 articles on groups during the period 1860–1880) and directed attention to a large number of difficult problems, introducing many fundamental concepts. For example, Jordan makes explicit the notions of isomorphism and homomorphism for (substitution) groups, introduces the term "solvable group" for the first time in a technical sense, introduces the concept of a composition series, and proves part of the Jordan-Hölder theorem, namely, that the indices in two composition series are the same (the concept of a quotient group was not explicitly recognized at this time); and he undertakes a very thorough study of transitivity and primitivity for permutation groups, obtaining results most of which have not since been superseded. Jordan also gives a proof that A_n is simple for $n > 4$.

An important part of the treatise is devoted to a study of the "linear group" and some of its subgroups. In modern terms these constitute the so-called classical groups, namely, the general linear group, the unimodular group, the orthogonal group, and the symplectic group. Jordan considers these groups only over finite fields, and proves their simplicity in certain cases. It should be noted, however, that he considers these groups as permutation groups rather than groups of matrices or linear transformations (see [29], [33]).

Jordan's *Traité* is a landmark in the evolution of group theory. His permutation-theoretic point of view, however, was soon to be overtaken by the conception of a group as a group of transformations (see (c) below). "The Traité marks a pause in the evolution and application of the permutation-theoretic group concept. It was an expression of Jordan's deep desire to effect a conceptual synthesis of the mathematics of his time. That he tried to achieve such a synthesis by relying on the concept of a permutation group, which the very next phase of mathematical development would show to have been unduly restricted, makes for both the glory and the limitations of the Traité . . ." (Wussing).

(b) *Abelian Groups*

As noted earlier, the main source for abelian group theory was number theory, beginning with Gauss' *Disquisitiones Arithmeticae*. In contrast to permutation theory, group-theoretic modes of thought in number theory remained implicit until about the last third of the 19th century. Until that time no explicit use of the term "group" was made, and there was no link to the contemporary, flourishing theory of permutation groups. We now give a sample of some implicit group–theoretic work in number theory, especially in algebraic number theory.

Algebraic number theory arose in connection with Fermat's conjecture concerning the equation $x^n + y^n = z^n$, Gauss' theory of binary quadratic forms, and higher reciprocity laws. Algebraic number fields and their arithmetical properties were the main objects of study. In 1846 G. L. Dirichlet studied the units in an algebraic number field and established that (*in our terminology*) the group of these units is a direct product of a finite cyclic group and a free abelian group of finite rank. At about the same time E. Kummer introduced his "ideal numbers," defined an equivalence relation on them, and derived, for cyclotomic fields, certain special properties of the number of equivalence classes (the so-called class number of a cyclotomic field; in our terminology, the order of the ideal class group of the cyclotomic field). Dirichlet had earlier made similar studies of quadratic fields.

In 1869 E. Schering, a former student of Gauss, investigated the structure of Gauss' (group of) equivalence classes of binary quadratic forms. He found certain fundamental classes from which all classes of forms could be obtained by composition. In group-theoretic terms, Schering found a basis for the abelian group of equivalence classes of binary quadratic forms.

L. Kronecker generalized Kummer's work on cyclotomic fields to arbitrary algebraic number fields. In a paper in 1870 on algebraic number theory, entitled "Auseinandersetzung einiger Eigenschaften der Klassenzahl idealer complexer Zahlen," he began by taking a very abstract point of view: he considered a finite set of arbitrary "elements," and defined an abstract operation on them which satisfied certain laws—laws which we may take nowadays as axioms for a finite abelian group:

Let θ', θ'' θ''',... be finitely many elements such that with any two of them we can associate a third by means of a definite procedure. Thus, if f denotes the procedure and θ', θ'' are two (possibly equal) elements, then there exists a θ''' equal to $f(\theta', \theta'')$ Furthermore, $f(\theta', \theta'') = f(\theta'' \; \theta')$, $f(\theta', f(\theta'', \theta''')) = f(f(\theta', \theta''), \theta''')$ and if θ'' is different from θ''' then $f(\theta', \theta'')$ is different

from $f(\theta', \theta''')$. Once this is assumed we can replace the operation $f(\theta', \theta'')$ by multiplication $\theta'.\theta''$ provided that instead of equality we employ equivalence. Thus using the usual equivalence symbol "~"we define the equivalence $\theta'. \; \theta''\sim\theta'''$ by means of the equation $f(\theta', \theta'') = \theta'''$.

Kronecker aimed at working out the laws of combination of "magnitudes," in the process giving an implicit definition of a finite abelian group. From the above abstract considerations Kronecker deduces the following consequences:

(i) If θ is any "element" of the set under discussion, then $\theta^k = 1$ for some positive integer k. If k is the smallest such then θ is said to "belong to k". If θ belongs to k and $\theta^m = 1$ then k divides m.

(ii) If an element θ belongs to k, then every divisor of k has an element belonging to it.

(iii) If θ and θ' belong to k and k' respectively, and k and k' are relatively prime, then $\theta\theta'$ belongs to kk'.

(iv) There exists a "fundamental system" of elements θ_1, θ_2, θ_3, ... such that the expression $\theta_1^{h_1}\theta_2^{h_2}\theta_3^{h_3}$ ... $(h_i = 1, 2, 3,..., n_i)$ represents each element of the given set of elements just once. The numbers n_1, n_2, n_3, ... to which, respectively θ_1, θ_2, θ_3, ... belong, are such that each is divisible by its successor; the product n_1, n_2, n_3,... equal to the totality of elements of the set.

The above can, of course, be interpreted as well known results on finite abelian groups; in particular (iv) can be taken as the basis theorem for such groups. Once Kronecker establishes this general framework, he applies it to the special cases of equivalence classes of binary quadratic forms and to ideal classes. He notes that when applying (iv) to the former one obtains Schering's result.

Although Kronecker did not relate his implicit definition of a finite abelian group to the (by that time) well established concept of a permutation group, of which he was well aware, he clearly recognized the advantages of the abstract point of view which he adopted:

The very simple principles . . . are applied not only in the context indicated but also frequently, elsewhere—even in the elementary parts of number theory. This shows, and it is otherwise easy to see, that these principles belong to a more general and more abstract realm of ideas. It is therefore proper to free their development from all inessential restrictions, thus making it unnecessary to repeat the same argument when applying it in different cases. . . . Also, when stated with all admissible generality, the presentation gains in simplicity and, since only the truly essential features are thrown into relief, in transparency.

The above lines of development were capped in 1879 by an important paper of G. Frobenius and L. Stickelberger entitled "On groups of commuting elements."Although Frobenius and Stickelberger built on Kronecker's work, they used the concept of an abelian group explicitly and, moreover, made the important advance of recognizing that the abstract group concept embraces 'congruences and Gauss' composition of forms as well as the substitution groups of Galois. (They

also mention, in footnotes, groups of infinite order, namely groups of units of number fields and the group of all roots of unity.) One of their main results is a proof of the basis theorem for finite abelian groups, including a proof of the uniqueness of decomposition. It is interesting to compare their explicit, "modem, formulation of the theorem to that of Kronecker ((iv) above):

A group that is not irreducible [indecomposable] can be decomposed into purely irreducible factors. As a rule, such a decomposition can be accomplished in many ways. However, regardless of the way in which it is carried out, the number of irreducible factors is always the same and the factors in the two decompositions can be so paired off that the corresponding factors have the same order.

They go on to identify the "irreducible factors" as cyclic groups of prime power orders. They then apply their results to groups of integers modulo m, binary quadratic forms, and ideal classes in algebraic number fields.

Georg Frobenius

The paper by Frobenius and Stickelberger is "a remarkable piece of work, building up an independent theory of finite abelian groups on its own foundation in a way close to modern views" (Fuchs).

(c) *Transformation Groups*

As in number theory, so in geometry and analysis, group-theoretic ideas remained implicit until the last third of the 19th century. Moreover, Klein's (and Lie's) explicit use of groups in geometry influenced conceptually rather than, technically the evolution of group theory, for it signified a genuine shift in the development of that theory from a preoccupation with permutation groups to the study of groups of transformations. (That is not to imply, of course, that permutation groups were no longer studied.) This transition was also notable in that it pointed to a turn from finite groups to infinite groups.

Klein noted the connection of his work with permutation groups but also realized the departure he was making. He stated that what Galois theory and his own program have in common is the investigation of "groups of changes," but added that "to be sure, the objects the changes apply to are different: there [Galois theory] one deals with a finite number of discrete elements, whereas here one deals with an infinite number of elements of a continuous manifold." To continue the analogy, Klein notes that just as there is a theory of permutation groups, "we insist on a theory of transformations, a study of groups generated by transformations of a given type."

Klein shunned the abstract point of view in group theory, and even his technical definition of a (transformation) group is deficient: "Now let there be given a sequence of transformations *A, B, C, . . .* If this sequence has the property that the composite of any two of its transformations yields a transformation that again belongs to the sequence, then the latter will be called a group of transformations." His work, however, broadened considerably the conception of a group and its applicability in other fields of mathematics. Klein did much to promote the view that group-theoretic ideas are fundamental in mathematics: "Group theory appears as a distinct discipline throughout the whole of modern mathematics. It permeates the most varied areas as an ordering and classifying principle."

There was another context in which groups were associated with geometry, namely, "motion−geometry;" i.e., the use of motions or transformations of geometric objects as group elements. Already in 1856 W. R. Hamilton considered (implicitly) "groups" of the regular solids. Jordan, in 1868, dealt with the classification of all subgroups of the group of motions of Euclidean 3-space. And Klein in his Lectures on the Icosahedron of 1884 "solved" the quintic equation by means of the symmetry group of the icosahedron. He thus discovered a deep connection between the groups of rotations of the regular solids, polynomial equations, and complex function theory. (In these Lectures there also appears the "Klein 4-group.")

Already in the late 1860's Klein and Lie had undertaken, jointly, "to investigate geometric or analytic objects that are transformed into themselves by *groups of changes*." (This is Klein's retrospective description, in 1894, of their program.) While Klein concentrated on discrete groups, Lie studied continuous transformation groups. Lie realized that the theory of continuous transformation groups was a very powerful tool in geometry and differential equations and he set himself the task of "determining sail groups of . . . [continuous] transformations." He achieved his objective by the early 1880's with the classification of these groups. A classification of discontinuous transformation groups was obtained by Poincaré and Klein a few years earlier.

Beyond the technical accomplishments in the areas of discontinuous and continuous transformation groups (extensive theories developed in both areas and both are still nowadays active fields of research), what is important for us in the founding of these theories is that

(i) They provided a major extension of the scope of the concept of a group—from permutation groups and abelian groups to transformation groups;

(ii) They provided important examples of infinite groups—previously the only objects of study were finite groups;

(iii) They greatly extended the range of applications of the group concept to include number theory, the theory of algebraic equations, geometry, the theory of differential equations (both ordinary and partial), and function theory (automorphic functions, complex functions).

All this occurred prior to the emergence of the abstract group concept. In fact, these developments were instrumental in the emergence of the concept of an abstract group, which we describe next.

Emergence of Abstraction in Group Theory

The abstract point of view in group theory emerged slowly. It took over one hundred years 'from the time of Lagrange's implicit group-theoretic work of 1770 for the abstract group concept to evolve. E. T. Bell discerns several stages in this process of evolution towards abstraction and axiomatization:

> The entire development required about a century. Its progress is typical of the evolution of any major mathematical discipline of the recent period; first, the discovery of isolated phenomena, then the recognition

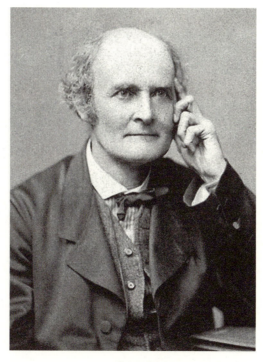

Arthur Cayley

of certain features common to all, next the search for further instances, (their detailed calculation and classification; then the emergence of general principles making further calculations, unless needed for some definite application, superfluous; and last, the formulation of postulates crystallizing in abstract form the structure of the system investigated.

Although somewhat oversimplified (as all such generalizations tend to be), this is nevertheless a useful framework. Indeed, in the case of group theory, first came the "isolated phenomena"— e.g., permutations, binary quadratic forms, roots of unity; then the recognition of "common features"—the concept of a finite group, encompassing both permutation groups and finite abelian groups (cf. the paper of Frobenius and Stickelberger cited in section 2(b)); next the search for "other instances"—in our case transformation groups (see section 2(c)); and finally the formulation of "postulates"—in this case the postulates of a group, encompassing both the finite and infinite cases. We now consider when and how the intermediate and final stages of abstraction occurred.

In 1854 Cayley, in a paper entitled "On the theory of groups, as depending on the symbolic equation $\theta^n = 1$," gave the first abstract definition of a finite group. (In 1858 R. Dedekind, in lectures on Galois theory at Gottingen, gave another.) Here is Cayley's definition:

A set of symbols 1, α, β,... all of them different, and such that the product of any two of them (no matter in what order), or the product of any one of them into itself, belongs to the set, is said to be a group.

Cayley goes on to say that

These symbols are not in general convertible [commutative], but are associative,

and

it follows that if the entire group is multiplied by any one of the symbols, either as further or nearer factor [i.e., on the left or on the right], the effect is simply to reproduce the group.

Cayley then presents several examples of groups, such as the quaternions (under addition), invertible matrices (under multiplication), permutations, Gauss' quadratic forms, and groups arising in elliptic function theory. Next he shows that every abstract group is (in our terminology) isomorphic to a permutation group, a result now known as "Cayley's theorem." He seems to have been well aware of the concept of isomorphic groups, although he does not define it explicitly. He introduces, however, the multiplication table of a (finite) group and asserts that an abstract group is determined by its multiplication table. He then goes on to determine all the groups of orders four and six, showing there are two of each by displaying multiplication tables. Moreover, he notes that the cyclic group of order n "is in every respect analogous to the system of the roots of the ordinary equation $x^n - 1 = 0$," and that there exists only one group of a given prime order.

Heinrich Weber

Cayley's orientation towards an abstract view of groups—a remarkable accomplishment at this time of the evolution of group theory—was due, at least in part, to his contact with the abstract work of G. Boole. The concern with the abstract foundations of mathematics was characteristic of the circles around Boole, Cayley, and Sylvester already in the 1840's. Cayley's achievement was, however, only a personal triumph. His abstract definition of a group attracted no attention at the time, even though Cayley was already well known. The mathematical community was apparently not ready for such abstraction: permutation groups were the only groups under serious investigation, and more generally, the formal approach to mathematics was still in its infancy. As M. Kline put it in his inimitable way: "Premature abstraction falls on deaf ears, whether they belong to mathematicians or to students."

It was only a quarter of a century later that the abstract group concept began to take hold. And it was Cayley again who in four short papers on group theory written in 1878 returned to the abstract point of view he adopted in 1854. Here he stated the general problem of finding all groups of a given order and showed that any (finite) group is isomorphic to a group of permutations. But, as he remarked, this "does not in any wise show that the best or easiest mode of treating the general problem is thus to regard it as a problem of substitutions; and it seems clear that the better course is to consider the general problem in itself, and to deduce from it the theory of groups of substitutions." These papers of Cayley, unlike those of 1854, inspired a. number of fundamental group-theoretic works.

Another mathematician who advanced the abstract point of view in group theory (and more generally in algebra) was H. Weber. It is of interest to see his "modern" definition of an abstract (finite) group given in a paper of 1882 on quadratic forms:

A system G of h arbitrary elements θ_1, θ_2, θ_h is called a group of degree h if it satisfies the following conditions:

I.   By some rule which is designated as composition or multiplication, from any two elements of the same system one derives a new element of the same system. In symbols $\theta_r \theta_s = \theta_t$

II.   It is always true that $(\theta_r \theta_s)\, \theta_t = \theta_r\,(\theta_s\,\theta_t) = \theta_r\,\theta_s\,\theta_t$

III.   From $\theta\theta_r = \theta\theta_s$, or from $\theta_r\,\theta = \theta_s\theta$ it follows that $\theta_r = \theta_s$.

Weber's and other definitions of abstract groups given at the time applied to *finite* groups only. They thus encompassed the two theories of permutation groups and (finite) abelian groups, which derived from the two sources of classical algebra (polynomial equations) and number theory, respectively. Infinite groups, which derived from the theories of (discontinuous and continuous) transformation groups, were not subsumed under those definitions. It was W. von Dyck who, in an important and influential paper in 1882 entitled "Group-theoretic studies," consciously included and combined, for the first time, all of the major historical roots of abstract group theory—the algebraic, number theoretic, geometric, and analytic. In von Dyck's own words:

> The aim of the following investigations is to continue the study of the properties of a group in its abstract formulation. In particular, this will pose the question of the extent to which these properties have an invariant character present in all the different realizations of the group, and the question of what leads to the exact determination of their essential group-theoretic content.

Von Dyck's definition of an abstract group, which included both the finite and infinite cases, was given in terms of generators (he calls them "operations") and defining relations (the definition is somewhat long). He stresses that "in this way all . . . *isomorphic* groups are included in a *single* group," and that "the *essence* of a group is no longer expressed by a particular presentation form of its operations but rather by their mutual relations." He then goes on to construct the free group on n generators, and shows (essentially, without using the terminology) that every finitely generated group is a quotient group of a free group of finite rank. What is important from the point of view of postulates for group theory is that von Dyck was the first to require explicitly the existence of an inverse in his definition of a group: "We require for our considerations that a group which contains the operation T_k must also contain its inverse T_k^{-1}." In a second paper (in 1883) von Dyck applied his abstract development of group theory to permutation groups, finite rotation groups (symmetries of polyhedra), number theoretic groups, and transformation groups.

Although various postulates for groups appeared in the mathematical literature for the next twenty years, the abstract point of view in group theory was not universally applauded. In particular, Klein, one of the major contributors to the development of group theory, thought that the "abstract formulation is excellent for the working out of proofs but it does not help one find new ideas and methods," adding that "in general, the disadvantage of the [abstract] method is that it fails to encourage thought."

Despite Klein's reservations, the mathematical community was at this time (early 1880's) receptive to the abstract formulations (cf. the response to Cayley's definition of 1854). The major reasons for this receptivity were:

(i)   There were now several major "concrete" theories of groups—permutation groups, abelian groups, discontinuous transformation groups (the finite and infinite cases), and continuous transformation groups, and this warranted abstracting their essential features.

(ii) Groups came to play a central role in diverse fields of mathematics, such as different parts of algebra, geometry,

Henri Poincare

number theory and several areas of analysis, and the abstract view of groups was thought to clarify what was essential for such applications and to offer opportunities for further applications.

(iii) The formal approach, aided by the penetration into mathematics of set theory and mathematical logic, became prevalent in other fields of mathematics, for example, various areas of geometry and analysis.

In the next section we will follow, very briefly, the evolution of that abstract point of view in group theory.

Consolidation of the Abstract Group Concept; *Dawn of Abstract Group* Theory

The abstract group concept spread rapidly during the 1880's and 1890's, although there still appeared a great many papers in the areas of permutation and transformation groups. The abstract viewpoint was manifested in two ways:

(a) Concepts and results introduced and proved in the setting of "concrete" groups were now reformulated and reproved in an abstract setting;

(b) Studies originating in, and based on, an abstract setting began to appear.

An interesting example of the former case is a reproving by Frobenius, in an abstract setting, of Sylow's theorem, which was proved by Sylow in 1872 for permutation groups. This was done in 1887, in a paper entitled "Neuer Beweis Sylowschen Satzes." Although Frobenius admits that the fact that every finite group can be represented by a group of permutations proves that Sylow's theorem must hold for all finite groups, he nevertheless wishes to establish the theorem abstractly: "Since the symmetric group, which is introduced in all these proofs, is totally alien to the context of Sylow's theorem, I have tried to find a new derivation of it . . . ."

Hölder was an important contributor to abstract group theory, and was responsible for introducing a number of group-theoretic concepts abstractly. For example, in 1889 he introduced the abstract notion of a quotient group (the "quotient group" was first seen as the Galois group of the "auxiliary equation," later as a homomorphic image and only in Hölder's time as a group of cosets), and "completed" the proof of the Jordan-Holder theorem, namely, that the quotient groups in a composition series are invariant up to isomorphism (see section 2(a) for Jordan's contribution). In 1893, in a paper on groups of order p^3, pq^2, pqr, and p^4, he introduced abstractly the concept of an automorphism of a group. Hölder was also the first to study simple groups abstractly. (Previously they were considered in concrete cases—as permutation groups, transformation groups, and so on.) As he says, "It would be of the greatest interest if a survey of all simple groups with a finite number of operations could be known." (By "operations" Hölder meant elements.) He then goes on to determine the simple groups of order up to 200.

Other typical examples of studies in an abstract setting are the papers by Dedekind and G. A. Miller in 1897/1898 on Hamiltonian groups— i.e., nonabelian groups in which all subgroups are normal. They (independently) characterize such groups abstractly, and introduce in the process the notions of the commutator of two elements and the commutator subgroup (Jordan had previously introduced the notion of commutator of two permutations).

The theory of group characters and the representation theory for finite groups (created at the end of the 19th century by Frobenius and Burnside/Frobenius/Molien, respectively) also belong to the area of abstract group theory, as they were used to prove important results about abstract groups.

Although the abstract group *concept* was well established by the end of the 19th century, "this was not accompanied by a general acceptance of the associated method of presentation in papers, textbooks, monographs, and lectures. Group-theoretic monographs based on the abstract group concept did not appear until the beginning of the 20th century. Their appearance marked the birth of abstract group theory" (Wussing).

The earliest monograph devoted entirely to abstract group theory was the book by J. A. de Seguier of 1904 entitled *Elements of the Theory of Abstract Groups*. At the very beginning of the book there is a set-theoretic introduction based on the work of Cantor: "De Seguier may have been the first algebraist to take note of Cantor's discovery of uncountable cardinalities" (B. Chandler and W. Magnus). Next is the introduction of the concept of a semigroup with two-sided cancellation law and a proof that a finite semigroup is a group. There is also a proof, by means of counterexamples, of the independence of the group postulates. De Séguier's book also includes a discussion of isomorphisms, homomorphisms, automorphisms, decomposition of groups into direct products, the Jordan-Hölder theorem, the first isomorphism theorem, abelian groups including the basis theorem, Hamiltonian groups, and finally, the theory of p-groups. All this is done in the abstract, with "concrete" groups relegated to an appendix. "The style of de Séguier is in sharp contrast to that of Dyck. There are no intuitive considerations . . . and there is a tendency to be as abstract and as general as possible . . ." (Chandler and Magnus).

De Seguier's book was devoted largely to finite groups. The first abstract monograph on group theory which dealt with groups in general, relegating finite groups to special chapters, was O. Schmidt's *Abstract Theory of Groups* of 1916. Schmidt, founder of the Russian school of group theory, devotes the first four chapters of his book to gyroup properties common to finite and infinite groups. Discussion of finite groups is postponed to chapter 5, there being ten chapters in all.

Divergence of Developments in Group Theory

Group theory evolved from several different sources, giving rise to various concrete theories. These theories developed independently, some for over one hundred years (beginning in 1770) before they converged (early 1880's) within the abstract group concept. Abstract group theory emerged and was consolidated in the next thirty to forty years. At the end of that period (around 1920) one can discern the divergence of group theory into several distinct "theories." Here is the

barest indication of some of these advances and new directions in group theory, beginning in the 1920's (with contributors and approximate dates):

(a) Finite group theory. The major problem here, already formulated by Cayley (1870's) and studied by Jordan and Hölder, was to find all finite groups of a given order. The problem proved too difficult and mathematicians turned to special cases (suggested especially by Galois theory): to find all simple or all solvable groups (cf. the Feit-Thompson theorem of 1963, and the classification of all finite simple groups in 1981).

(b) Extensions of certain results from finite group theory to infinite groups with finiteness conditions; e.g., O. J. Schmidt's proof, in 1928, of the Remalc-Krull-Schmidt theorem.

(c) Group presentations (Combinatorial Group Theory), begun by von Dyck in 1882, and continued in the 20th century by M. Dehn, H. Tietze, J. Nielsen, E. Artin, O. Schreier, et al.

(d) Infinite abelian group theory (H. Prüfer, R. Baer, H. Ulm et al.—1920's to 1930's).

(e) Schreier's theory of group extensions (1926), leading later to the cohomology of groups.

(f) Algebraic groups (A. Borel, C. Chevalley et al.—1940's).

(g) Topological groups, including the extension of group representation theory to continuous groups (Schreier, E. Cartan. L. Pontrjagin, I. Gelfand, J. von Neumann et al.—1920's and 1930's).

FIGURE 1 gives a diagrammatic sketch of the evolution of group theory as outlined in the various sections and as summarized at the beginning of this section.

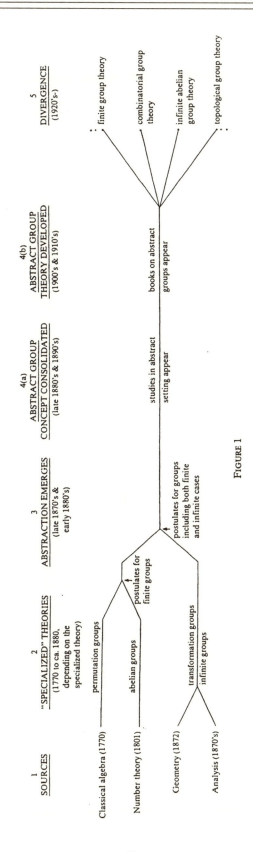

FIGURE 1

BIBLIOGRAPHY

R. G. Ayoub, "Paolo Ruffini's contributions to the quintic," *Arch. Hist. Ex. Sc.*, 23 (1980) 253–277.

*E. T. Bell, *The Development of Mathematics*, McGraw Hill, 1945.

G. Birkhoff, "Current trends in algebra," *Amer. Math. Monthly*, 80 (1973) 760–782 and 81 (1974) 746.

———. "The rise of modern algebra to 1936," *Men and Institutions in American Mathematics*, eds. D. Tarwater, J. T. White and J. D. Miller, Texas Tech Press, 1976, pp. 41–63.

N. Bourbaki, *Eléments d'Histoire des Mathématiques*, Hermann, 1969.

J. E. Burns, "The foundation period in the history of group theory," *Amer. Math. Monthly*, 20 (1913) 141–148.

B. Chandler and W. Magnus, *The History of Combinatorial Group Theory: A Case Study in the History of Ideas*, Springer-Verlag, 1982.

A. Dahan, "Les travaux de Cauchy sur les substitutions. Etude de son approche du concept de groupe," *Arch. Hist. Ex. Sc.*, 23 (1980) 279–319.

J. Dieudonné (ed.), *Abrégé d'Histoire des Mathématiques*, 1700–1900, 2 vols., Hermann, 1978.

P. Dubreil, "L'algèbre, en France, de 1900 à 1935," *Cahiers du seminaire d'histoire des mathématiques*, 3 (1981) 69–81.

C. H. Edwards, *The Historical Development of the Calculus*, Springer-Verlag, 1979.

H. M. Edwards, *Galois Theory*, Springer-Verlag, 1984.

J. A. Gallian, The search for finite simple groups, *Mathematics Magazine*, 49 (1976) 163–179.

D. Gorenstein, *Finite Simple Groups: An Introduction to Their Classification*, Plenum Press, 1982.

———. *The Classification of Finite Simple Groups*, Plenum Press, 1983.

R. R. Hamburg, "The theory of equations in the 18th century: The work of Joseph Lagrange," *Arch. Hist. Ex. Sc.*, 16 (1976/77) 17–36.

T. Hawkins, "Hypercomplex numbers, Lie groups, and the creation of group representation theory," *Arch. Hist. Ex. Sc.*, 8 (1971/72) 243–287.

———. "The *Erlanger Programme* of Felix Klein: Reflections on its place in the history of mathematics," *Hist. Math.*, 11 (1984) 442–470.

B. M. Kiernan, "The development of Galois theory from Lagrange to Artin," *Arch. Hist. Ex. Sc.*, 8 (1971/72) 40–154.

F. Klein, "Development of Mathematics in the 19th Century (transl. from the 1928 German ed. by M. Ackerman), in Lie Groups: History, Frontiers and Applications," vol. IX, ed. R. Hermann, *Math. Sci. Press*, 1979, 1–361.

M. Klein, *Mathematical Thought from Ancient to Modern Times*, Oxford Univ. Press, 1972.

D. R. Lichtenberg, "The Emergence of Structure in Algebra," Doctoral Dissertation, Univ. of Wisconsin, 1966.

U. Merzbach, "Development of Modern Algebraic Concepts from Leibniz to Dedekind," Doctoral Dissertation, Harvard Univ., 1964.

G. A. Miller, *History of the theory of groups, Collected Works*, 3 vols. 427–467, 1–18, 1–15, Univ. of Illinois Press, 1935, 1938, 1946.

L. Novy, *Origins of Modern Algebra*, Noordhoff, 1973.

O. J. Schmidt, *Abstract Theory of Groups*, W. H. Freeman & Co., 1966. (Translation by F. Holling and J. B. Roberts of the 1916 Russian edition.)

*Available as a Dover reprint.

J.-A. de Séguier, *Théorie des Groupes Finis. Eléments de la Théorie des Groupes Abstraits,* Gauthier Villars, Paris, 1904.

L. A. Shemetkov, "Two directions in the development of the theory of non-simple finite groups," *Russ. Math. Surv.* 30 (1975) 185–206.

R. Silvestri, "Simple groups of finite order in the nineteenth century," *Arch. Hist. Ex. Sc.,* 20 (1979) 313–356.

J. Tarwater, J. T. White, C. Hall, and M. E. Moore (eds.), *American Mathematical Heritage: Algebra and Applied Mathematics,* Texas Tech. Press, 1981. Has articles by Feit, Fuchs, and MacLane on the history of finite groups, abelian groups, and abstract algebra, respectively.

B. L. Van der Waerden, "Die Algebra seit Galois," *Jahresbericht der Deutsch Math. Ver.,* 68 (1966) 155–165.

W. C. Waterhouse, "The early proofs of Sylow's theorem," *Arch. Hist. Ex. Sc.,* 21 (1979/80) 279–290.

*H. Wussing, *The Genesis of the Abstract Group Concept,* MIT Press, 1984. (Translation by A. Shenitzer of the 1969 German edition.)

*Available as a Dover reprint.

7

The Men Responsible for the Development of Vectors

GEORGE J. PAWLIKOWSKI

*I*N ORDER to fully appreciate the study of vectors, one should have some idea of how they came about. This paper is devoted to the men responsible for the development of vectors. Since not all the men responsible for vector development could be discussed here, I have taken a few of the important men in the field and searched for information about them. The following is the result of my study.

Simon Stevin

Simon Stevin was born in Bruges, Belgium, in 1548. He was an engineer, physicist, and mathematician who combined practical sense, theoretical understanding, and originality. His physics led him to the discovery of the hydrostatic paradox. He introduced decimal fractions into arithmetic in *La disme* in 1585. This led to a great improvement in the system of measurement. His most famous work was *De Beghinselen der Weeghconst*. It was in this work, which came out in 1586, that he concerned himself with statics. Stevin investigated statics and determined accurately the force necessary to sustain a body on a plane inclined at any angle to the horizon. He was the first scientist since Archimedes to work with statics and vectors.

Reprinted from *Mathematics Teacher* 60 (Apr., 1967): 393–96; with permission of the National Council of Teachers of Mathematics.

Sir William R. Hamilton

William R. Hamilton was born in Dublin, Ireland, on August 4, 1805. At thirteen, he started to become interested in neutron readings. He studied about the calculus of directed line segments in space and the meaning assigned to their products. By doing this, Hamilton used vector analysis in his study of differential geometry. He studied at Trinity College in Dublin and, when he graduated in 1827, he was appointed as the professor of astronomy.

In 1835, he introduced couples of moments and he used couples of real numbers (a, b). In 1843, he had a thought that led to the birth of the quaternions. He announced this thought at a meeting of the Royal Irish Academy. Hamilton is remembered most for the quaternions. He developed the quaternions along the lines of quaternion differentials and the linear vector function. The importance of these quaternions is due to their extension of the concept of number. The geometric application of the quaternions is also important. Hamilton expected and anticipated the quaternions would be found useful in physics for giving a mathematical explanation of the universe. However, this was never realized. Hamilton claimed that negative numbers and imaginary numbers were to be treated in the science of order and not in the science of magnitude.

A paper which he presented in 1848 laid the groundwork for linear algebra. In this same paper he introduced n-tuples and defined noncom-mutative multiplication for some aspects of what we know today as vectors. In 1853, he published his lectures on the quaternions, and his *Elements of Quaternions* was published in 1866, just one year after his death. This publication dealt with the sum of a scalar and a vector and with linear vector functions. Much of Hamilton's work is still unpublished. He was a brilliant man, fluent in some twelve languages. He used Newton's equations to state his principle which, since it generalizes the principle of least time, is basic to dynamic theory. Hamilton also worked extensively with the theory of fifth-degree algebraic equations.

Hermann Günther Grassmann

Hermann Grassmann was born in Stettin on April 15, 1809. He made his home in Stettin and taught at the gymnasium. Grassmann started with an algebra of points and went on develop his system and theories on vector algebra and the theory of a line vector. In 1844, he wrote *Lineale Ausdehnungslehre* in Euclidean form. In this book, Grassmann developed and worked with a geometry of n-dimensions. He used invariant symbolism which we now recognize as a vector. Even though this was quite abstract in his time, it has become an important tool for the twentieth-century mathematician and scientist. He loved mathematics and did much research in it.

Georg Friedrich Bernhard Riemann

Bernhard Riemann was born in the village of Breselenz, Hanover, Germany, on September 17, 1826. He received his early education from his father, who was a Lutheran pastor. His father wanted him to enter the ministry, but Bernhard's love for mathematics won out. He was educated at Göttingen and Berlin under Gauss and Jacobi. He developed a form of non-Euclidean geometry based on a postulate which does not permit parallel lines. His geometry is too complicated to discuss here;

however, it may be said that the parallel displacement of a vector has some of the properties of parallelism that are familiar from elementary Euclidean geometry. Much of Riemann's geometry is concerned with vectors of this nature. His *Gesammelte Mathematische Werke* was posthumously published in 1876. Riemann is known for his theories on differential geometry, contributions to the theory of functions of a complex variable, and the Riemann definite integral—which is the basic form as presented in calculus today. Einstein used much of Riemann's geometry in his work on the theory of relativity. Riemann was a member of the Royal Society. He was a sick man most of his life. He died at an early age on July 20, 1866, of tuberculosis.

Elwin Bruno Christoffel

Elwin Christoffel was a German mathematician who was born in 1829 and died in 1900. He did quite a bit of work with tensor analysis, which is a generalization of vector analysis. In 1870, along with R. Lipschitz, he wrote two papers that dealt with Riemann's theory of quadratic differential forms. It was in these papers that Christoffel introduced the symbols bearing his name. These Christoffel symbols find use in tensor analysis in connection with curvatures of the coordinate system. This type of symbolism was just right for the transformation theory of quadratic differential forms.

Peter Guthrie Tait

Peter Tait was born in Dalkeith, Scotland, on April 28, 1831. He was a mathematician and a physicist who did much research on Sir William R. Hamilton's quaternions. His research material was published in *Elementary Treatise on Quaternions*. Tait also made use of vector analysis in differential geometry and the study of his golf-ball problem. However, his important publications were on the foundation of the kinetic theory of gases. Tait died in the year of 1901.

Josiah Willard Gibbs

J. Willard Gibbs was born in New Haven, Connecticut, on February 11, 1839. His father taught theology at the Yale Divinity School. Gibbs became a student at Yale and stayed there until he received his Ph.D. degree. He was a brilliant student, not only in mathematics, but in the humanities as well. When his studies at Yale terminated, he went abroad to study. In Europe he became acquainted with the world's leading mathematicians. In 1871, he returned to New Haven and started teaching mathematical physics at Yale.

Starting around 1880, he worked to develop geometric algebra into a system of vector algebra suitable for the needs of mathematical physicists. Soon after this date, he privately printed *Elements of Vector Analysis* for his pupils. He was always reluctant about publishing a book, but finally, in 1901, E. B. Wilson's *Vector Analysis* book was published. Dr. Wilson was a student of Gibbs and wrote the book using notes taken from Gibbs's lectures. Gibbs also did work with optics, statistical mechanics, and physical chemistry. He made and used diagrams and models to illustrate his theories. Gibbs adopted the best and simplest system for vectors from Hamilton's work on quaternions and Grassmann's work on geometric algebra. He used the above material in his lecture notes. *The Collected Works of J. Willard Gibbs* (1928) contains all of his published writings. He was a fellow of the Royal Society and a member of the National Academy of Sciences. He died in New Haven, April 28, 1903.

William Kingdon Clifford

William Clifford was born at Exeter, England, in 1845. He was a mathematician and a philosopher. Clifford believed that applied geometry is a part of physics and not a part of pure mathematics. His classification of loci, one year before his death, was an introduction to the study of *n*-dimensional space in a direction mainly projective. The methods employed in this study have been those of analytic, synthetic, and differential geometry, as well as vector analysis. He was a teacher at Trinity College in Cambridge and at University College in London. Clifford was the first Englishman who actually understood Riemann. With Riemann, he shared a deep interest in the origin of space conceptions. He also developed a geometry of motion for the study of which he generalized Hamilton's quaternions into what he called biquaternions. He wrote *The Common Sense of the Exact Sciences* and a paper titled "On the space theory of matter." This paper came out forty years before Einstein announced his theory of gravitation. Clifford died of tuberculosis in Madena in 1879.

Oliver Heaviside

Oliver Heaviside was born in Camden Town, England. He was a physicist and an electrical engineer who did much with long-distance telephone communication. There are many mathematicians who opposed Heaviside's work because of his methods. His theory was excellent, but he lacked a rigorous mathematical foundation for his work. His *Electrical Papers*, in 1892, and his *Electromagnetic Theory* (1893–1912) are important in the development of the electric theory. Heaviside worked with vectors in England about the same time that Gibbs worked with them in the United States. His work on vectors was a combination of Hamilton's and Tait's, which he altered and adapted to meet his own requirements. As a result of the above, he had a type of vector algebra much like that of Gibbs. Oliver Heaviside was a fellow of the Royal Society. He died in 1925.

Gregorio Ricci-Curbastro

Gregorio Ricci-Curbastro was an Italian mathematician who was born in 1853 and died in 1925. He read the papers that were written by E. Christoffel and R. Lipschitz in 1870 on Riemann's theory of quadratic differential forms.

These papers led Gregorio Ricci-Curbastro, after much study, to his absolute differential calculus. The symbols introduced in the papers by Christoffel and Lipschitz fitted nicely into Ricci-Curbastro's calculus. He helped in the development of the algebra and the calculus of tensors. The above-mentioned tensor analysis is a generalization of vector analysis.

Tullio Levi-Civita

Tullio Levi-Civita was born in 1873 and died in 1941. He studied under Ricci. Along with Ricci and fellow students, he helped to develop absolute differential calculus into the theory of tensors. Tensors provided a unification of many invariant symbolisms. Tullio Levi-Civita and Gergorio Ricci-Curbastro published a systematic account of the research of Ricci in 1901. Levi-Civita was a professor of mathematics at the University of Rome.

*Available as a Dover reprint.

BIBLIOGRAPHY

Bezuszka, Stanley J. *Co-op Unit Study Program, Course I.* Boston: College Mathematics Institute Press, 1960.

Cajori, F. *A History of Mathematics.* New York: Macmillan Co., 1919.

Collier's Encyclopedia. New York: Crowell-Collier Publishing Co., 1962.

Newman, James R. (ed.). *The Harper Encyclopedia of Science.* New York: Harper & Row, 1963.

*——— (ed.). *The World of Mathematics.* New York: Simon & Schuster, 1956.

*Smith, David E. *A Source Book in Mathematics.* New York: McGraw-Hill Book Co., 1929.

*Struik, Dirk J. *A Concise History of Mathematics.* New York: Dover Publications, 1948.

The Noncommutative Algebra of William Rowan Hamilton

By the early nineteenth century, the study of algebra had evolved from the mere manipulation of symbols to a more abstract investigation of the laws of mathematical operations and how they combined objects. William Rowan Hamilton (1805–1865) was a professor of mathematics at Trinity College, Dublin. In 1833, he devised an algebra for working with number couples. This algebra could be readily applied to complex numbers where a number of the form $a + bi$ could be represented by the ordered couple (a, b). Hamilton sought to extend his algebra further to the consideration of number triples; however, he ran into difficulty. In 1843, he finally extended his theory, but with number quadruples not triples. He called his groupings of four numbers "quaternions." The algebra of quaternions was most unusual—it was not commutative! "Multiplication" with Hamilton's system can be summarized as follows:

Let: $[1, 0, 0, 0]$ be represented by 1,
$[0, 1, 0, 0]$ by i,
$[0, 0, 1, 0]$ by j, and
$[0, 0, 0, 1]$ by k; then the properties of
"multiplication" are illustrated in the Cayley Table below:

*	1	i	j	k
1	1	i	j	k
i	i	-1	k	$-j$
j	j	$-k$	-1	i
k	k	j	$-i$	-1

Thus, the elements heading each column are combined with the elements leading each row to produce the result in the intersection of the column and row, i.e., $j * i = k$, etc.

8

Matrix Theory I: Arthur Cayley— Founder of Matrix Theory

RICHARD W. FELDMANN, JR.

ALTHOUGH THE TERM "matrix" was introduced into mathematical literature by James Joseph Sylvester[1] in 1850, the credit for founding the theory of matrices must be given to Arthur Cayley, since he published the first expository articles on the subject.

Arthur Cayley[2] was born in Richmond, England, on August 16, 1821. At the age of fourteen he entered King's College in London, and in 1838 he went to Trinity College, from which he graduated with highest honors. He left teaching a few years later, since a permanent position required the taking of religious vows. After practicing law to the extent that it supported him but did not interfere with his mathematical studies, he accepted the newly formed Sadlerian Professorship of Pure Mathematics at Cambridge in 1863. He remained in the capacity until his death on Jaunary 26, 1895, except for the first half of 1882, when he lectured at Johns Hopkins University at the request of Sylvester. Besides his work in matrix theory, Cayley also published copiously in the fields of analytic geometry of n dimensions, determinant theory, linear transformations, and skew surfaces. Arthur Cayley was a voluminous writer, whose collected papers fill twelve large volumes. His most productive period was between 1863 and 1883, when he published about 430 papers.

Reprinted from *Mathematics Teacher* 57 (Oct., 1962): 482–84; with permission of the National Council of Teachers of Mathematics.

Cayley's introductory paper in matrix theory was written in French and published in a German periodical.[3] In this paper, matrices are introduced to simplify the notation which arises in simultaneous linear equations. The set of equations

$$
\begin{aligned}
\xi &= \alpha x + \beta y + \gamma z + \cdots \\
\eta &= \alpha' x + \beta' y + \gamma' z + \cdots \\
\zeta &= \alpha'' x + \beta'' y + \gamma'' z + \cdots \\
\end{aligned}
$$

is written as

$$
\begin{aligned}
(\xi, \eta, \zeta, \cdots) \\
= (\ \alpha, \quad \beta, \quad \gamma, \quad \cdots) (x, y, z, \cdots). \\
\quad \alpha', \quad \beta', \quad \gamma' z, \cdots \\
\quad \alpha'', \quad \beta'', \quad \gamma'' z, \cdots \\
\end{aligned}
$$

The same article also introduces, although quite sketchily, the ideas of inverse matrix and of matrix multiplication, or "compounding" as Cayley called it.

The above basic properties are expanded in a second expository article[4] which also lists many additional properties of matrices. In this important paper, Cayley works mostly with square matrices with nine elements. He represents the zero matrix,

$$\begin{pmatrix} 0, & 0, & 0 \\ 0, & 0, & 0 \\ 0, & 0, & 0 \end{pmatrix},$$

by "0" and the "matrix unity,"

$$\begin{pmatrix} 1, & 0, & 0 \\ 0, & 1, & 0 \\ 0, & 0, & 1 \end{pmatrix},$$

by "1."

In introducing the algebra of matrices, Cayley defines the addition of two matrices by

$$\begin{pmatrix} a, & b, & c \\ a', & b', & c' \\ a'', & b'', & c'' \end{pmatrix} + \begin{pmatrix} \alpha, & \beta, & \gamma \\ \alpha', & \beta', & \gamma' \\ \alpha'', & \beta'', & \gamma'' \end{pmatrix}$$

$$= \begin{pmatrix} a + \alpha, & b + \beta, & c + \gamma \\ a' + \alpha', & b' + \beta', & c' + \gamma' \\ a'' + \alpha'', & b'' + \beta'', & c'' + \gamma'' \end{pmatrix}$$

and states, without proof, that matrices are commutative[5] and associative under addition. Two types of multiplication are exhibited. The first is designated as "scalar multiplication," that is

$$m \begin{pmatrix} a, & b, & c \\ a', & b', & c' \\ a'', & b'', & c'' \end{pmatrix} = \begin{pmatrix} ma, & mb, & mc \\ ma', & mb', & mc' \\ ma'', & mb'', & mc'' \end{pmatrix}.$$

The second type of multiplication is called "compounding," according to the following scheme:

noted that $M^n \cdot M^p = M^{n+p}$. Commutative and anticommutative[6] matrices under "compounding" are introduced, using the names of "convertible" and "skew-convertible" matrices.

In Section 17 of this same *Memoir*, Cayley exhibits the inverse of

$$\begin{pmatrix} a, & b, & c \\ a', & b', & c' \\ a'', & b'', & c'' \end{pmatrix}$$

in the form

$$\frac{1}{\nabla} \begin{pmatrix} \partial_a \nabla, & \partial_{a'} \nabla, & \partial_{a''} \nabla \\ \partial_b \nabla, & \partial_{b'} \nabla, & \partial_{b''} \nabla \\ \partial_c \nabla, & \partial_{c'} \nabla, & \partial_{c''} \nabla \end{pmatrix},$$

where ∇ is the determinant of the matrix and $\partial_x \nabla$ is the determinant obtained from ∇ by replacing the element x in ∇ by 1 and all other elements in the row and column containing x by zeros, which, in effect, makes $\partial_x \nabla$ the cofactor of x. When ∇ is zero, the matrix is called "indeterminate"[7] and has no inverse.

Cayley states, "It may be added that the matrix zero is indeterminate; and that the product of two matrices may be zero without either of the factors being zero, if only the matrices are one or both of them indeterminate." This statement is erroneous in that both matrices must be indeterminate.[8]

$$\begin{pmatrix} a, & b, & c \\ a', & b', & c' \\ a'', & b'', & c'' \end{pmatrix} \begin{pmatrix} \alpha, & \beta, & \gamma \\ \alpha', & \beta', & \gamma' \\ \alpha'', & \beta'', & \gamma'' \end{pmatrix}$$

$$= \begin{pmatrix} (a, b, c \langle \alpha, \alpha', \alpha''), & (a, b, c \langle \beta, \beta', \beta''), & (a, b, c \langle \gamma, \gamma', \gamma'') \\ (a', b', c' \langle \alpha, \alpha', \alpha''), & (a', b', c' \langle \beta, \beta', \beta''), & (a', b', c' \langle \gamma, \gamma', \gamma'') \\ (a'', b'', c'' \langle \alpha, \alpha', \alpha''), & (a'', b'', c'' \langle \beta, \beta', \beta''), & (a'', b'', c'' \langle \gamma, \gamma', \gamma'') \end{pmatrix},$$

where $(X, Y, z \langle A, A', A'')$ is a symbol representing $XA + YA' + ZA''$. This form of multiplication is said to be associative, but in general, not commutative.

The nth power of a matrix M is defined as a matrix "compounded" by itself $n-1$ times. It is also

The transposed matrix is defined by the equation

$$\text{tr} \begin{pmatrix} a, & b \\ c, & d \end{pmatrix} = \begin{pmatrix} a, & c \\ b, & d \end{pmatrix}.$$

Cayley further discusses this matrix by stating that

$$\text{tr} (LMN) = (\text{tr } N) (\text{tr } M) (\text{tr } L).$$

Symmetric and skew-symmetric matrices are defined by tr $M = M$ and tr $M = -M$ respectively.

The remainder of the paper discusses the Cayley-Hamilton theorem[9] and methods for finding roots and powers of a matrix. These methods are so vague and cumbersome that no further mention of them will be made.

The final statement in the article refers to rectangular matrices. Cayley notes here that an n by m matrix[10] can be added only to an n by m matrix, and that an n by m matrix can be compounded only by an m by p matrix. He concludes with the observation that the transpose of an n by m matrix is an m by n matrix.

Working with the transformation $\phi(x) = (ax + b)/(cx + d)$ in 1880,[11] Cayley showed that $\phi^n(x)$ has for coefficients the elements of

$$\begin{pmatrix} a, & b \\ c, & d \end{pmatrix}^n.$$

He proved that

$$\begin{pmatrix} a, b \\ c, d \end{pmatrix}^n = \left(\frac{1}{\lambda^2 - 1} \right) \left(\frac{a + d}{\lambda + 1} \right)^{n-1}$$

$$\left\{ (\lambda^{n-1} - 1) \begin{pmatrix} a, & b \\ c, & d \end{pmatrix} + (\lambda^n - \lambda) \begin{pmatrix} -d, & b \\ c, & -a \end{pmatrix} \right\},$$

where λ is determined from

$$(\lambda + 1)^2 / \lambda = (a + d)^2/(ad - bc).$$

An article by A. Buchheim[12] relates that Cayley developed notation to differentiate between right-hand and left-hand division; ab^{-1} and $b^{-1}a$ were represented by

respectively. This notation does not appear in Cayley's published works, and probably comes from his correspondence.

NOTES

1. "Additions to the Articles in the September Number of This Journal 'On a New Class of Theorems ...' and 'On Pascal's Theorem'," *Philosophical Magazine*, series three, XXXVII (1850), 363–70. Sylvester used the term to represent a rectangular portion of a determinant.

2. More biographical details on Arthur Cayley's life can be found in E. T. Bell, *Men of Mathematics* (New York: 1937); Alexander MacFarlane, *Lectures on Ten British Mathematicians* (New York: 1916); G. B. Halsted, "Arthur Cayley," *American Mathematical Monthly*, II (1895), 102–06.

3. "Remarques sur la notation des fonctions algebraiques," Crelle's *Journal für reine und angewandte Mathematik*, L (1855), 282–85.

4. "Memoir on the theory of matrices," *Philosophical Transactions of the Royal Society of London*, CXLVIII (1858), 17–37.

5. Cayley used the term "convertible."

6. $MN = -NM.$

7. We now use the term *singular*.

8. Consider $AB = 0$, A indeterminate and B not indeterminate. Then $ABB^{-1} = 0$, and $A = 0$.

9. This will be discussed in article III of this series, "The Characteristic Equation."

10. An n by m matrix has n rows and m columns.

11. "On the Matrix $\begin{pmatrix} a, & b \\ c, & d \end{pmatrix}$ and in Connexion Therewith the Function as $(ax + b)/(cx + d)$," *Messenger of Mathematics*, IX (1880), 104–09.

12. A. Buchheim, "On the Theory of Matrices," *Proceedings of the London Mathematical Society*, XVI (1885), 63–82.

Matrix Theory II: Basic Properties

RICHARD W. FELDMANN, JR.

$\mathcal{A}$FTER CAYLEY PUBLISHED his expository articles on the theory of matrices, many mathematicians published papers expanding Cayley's ideas and developing new concepts, notations, and terminology. Some of these were fruitful and are still used, while others never turned up in print again.

Division

Georg Frobenius[1] proposed that the quotient A/B be used to represent multiplication of A by B^{-1} when $AB^{-1} = B^{-1}A$. He then showed that $(A/B)^{-1}$ is B/A. Since matrix multiplication is not, in general, commutative, a second German mathematician, Kurt Hensel[2], suggested $A\backslash B$ and B/A to represent $A^{-1}B$ and BA^{-1}, respectively. However, both notations were soon dropped and the word *division* dropped from usage in favor of *multiplication by an inverse*.

Absolute Value

The concept of absolute value[3] of a matrix is still seen in modern mathematical literature, but quite infrequently. It was first proposed in 1925 by J. H. M. Wedderburn, who defined the absolute value of a matrix $A = (a_{ij})$, represented by $\lfloor A \rceil$, to be the square root of

Reprinted from *Mathematics Teacher* 57 (Nov., 1962): 589–90; with permission of the National Council of Teachers of Mathematics.

$$\sum_{i,j=1}^{n} a_{ij} \bar{a}_{ij}$$

where $\bar{a}_{ij}$ is the complex conjugate of a_{ij}. Wedderburn developed some theorems on absolute values. He proved that

$$\lfloor A + B \rceil \leq \lfloor A \rceil + \lfloor B \rceil \text{ and } \lfloor AB \rceil \leq \lfloor A \rceil \lfloor B \rceil.$$

Representing a scalar by λ, he showed

$$\lfloor \lambda A \rceil = |\lambda| \lfloor A \rceil \text{ and } \lfloor \lambda I \rceil = n^{1/2} |\lambda|.$$

The author stated that this concept was foreshadowed in 1909 by I. Schur[4] when he observed that $\lfloor A \rceil^2$ is the trace of $A\bar{A}^T$, where $\bar{A}^T$ is the transpose of the matrix formed by replacing each element in A by its complex conjugate.

Check for Matrix Multiplication

A fairly simple check for accuracy in multiplying two square matrices, $A = (a_{ij})$ and $B = (b_{ij})$, together was published in 1924 by W. E. Roth.[5] Setting

$$a_i = \sum_{j=1}^{n} a_{ij} \qquad \alpha_j = \sum_{i=1}^{n} a_{ij}$$

$$b_i = \sum_{j=1}^{n} b_{ij} \qquad \beta_j = \sum_{i=1}^{n} b_{ij}$$

the sum of the elements in AB is

$$\sum_{k=1}^{n} \alpha_k b_k$$

and the sum of the elements in BA is

$$\sum_{k=1}^{n} a_k \beta_k.$$

This result is extended to show that the sum of the elements in A^r is equal to

$$\sum_{i=1}^{n} \sum_{j=1}^{n} a_{ij}^{r-2} \alpha_i a_j.$$

$$| M - rI | = (-1)^n r^n + a_1 r^{n-1} + a_2 r^{n-2} + \cdots + a_n$$

and

$$| A - rI | = (-1)^n r^n + b_1 r^{n-1} + b_2 r^{n-2} + \cdots + b_n,$$

then $\lim a_i = b_i$ for $i = 1, 2, \cdots, n$.

Terminology

As the number of published papers increased, James Joseph Sylvester (1814–1897) introduced a terminology[6] which he hoped would eliminate much of the verbiage necessary when discussing the position of an element in a matrix. He proposed in 1884 that the element in the mth row and nth column be referred to as the element of "latitude" m and "longitude" n.

Limit of a Matrix

Refining some ideas of Giuseppe Peano (1888) and G. A. Bliss (1905), an article[7] by H. B. Phillips in the *American Journal of Mathematics* defined the limit of an n by n matrix $M = (m_{ij})$, whose elements are functions, to be the matrix $A = (a_{ij})$, where $\lim m_{ij} = a_{ij}$. The coefficients in the characteristic function of M were shown to approach, in the limit, the corresponding coefficients in the characteristic function of A. Symbolically, if

REFERENCES

1. "Über lineare Substitutionen und bilineare Formen," Crelle's *Journal für reine und angewandte Mathematik*, LXXXIV (1878), 1–63.

2. "Über den Zusammenanhang zwischen den Systemen und ihren Determinanten," Crelle's *Journal für reine und angewandte Mathematik*, CLIX (1928), 246-54.

3. "The Absolute Value of the Product of Two Matrices," *Bulletin of the American Mathematical Society*, XXXI (1925), 304-8.

4. "Über die charakteristischen Wurzeln einer linear Substitution mit einer Anwendung auf die Theorie der Integralgleichungen," *Mathematische Annalen*, LXVI (1909), 488–510.

5. "A Convenient Check on the Accuracy of the Product of Two Matrices," *American Mathematical Monthly*, XXXVI (1929), 37–38.

6. "Lectures on the Principles of Universal Algebra," *American Journal of Mathematics*, VI (1884), 270–86.

7. "Functions of Matrices," *American Journal of Mathematics*, XLI (1919), 266–78.

10

Matrix Theory III: The Characteristic Equation; Minimal Polynomials

RICHARD W. FELDMANN, JR.

$\mathcal{T}$HE CHARACTERISTIC EQUATION of a matrix M is found by expanding the determinant of $M - xI$, where I is the identity matrix, and setting it equal to zero. The roots of this equation are the characteristic roots of M.

The first appearance of the characteristic equation in matrix theory was in 1858 in Cayley's basic paper.[1] He does not refer to the equation by this name and used it to prove the theorem now known as the Cayley-Hamilton theorem. However, the terms "characteristic equation" and "characteristic root" are borrowed from papers on determinant theory. The terms "latent equation" and "latent root" were introduced into the literature by J. J. Sylvester. This equation was once called, but now very infrequently, the "secular equation," in honor of Laplace and his research on the secular inequalities of the planets.[2]

A relationship between the coefficients of the characteristic equation and the sums of the determinants of the various-sized principal minors of the matrix was demonstrated by William H. Metzler[3] in 1892. He proved that if the equation is written

$$x^n - m_1 x^{n-1} + m_2 x^{n-2} - \cdots \pm m_{n-1} x \mp m_n = 0,$$

then the determinant of the matrix is m_n, and m_k is the sum of the principal k-rowed minors of the determinant of the matrix.

Thirty-six years later, a German mathematician, Kurt Hensel,[4] published a proof that a non-zero constant term is required for the existence of the inverse matrix, but this is subsumed under Metzler's findings of 1892.

A special case was treated by Georg Frobenius,[5] when he proved that the characteristic function of a nilpotent matrix, i.e., a matrix some positive integral power of which is the zero matrix, is of the form x^r.

Certain special types of matrices were found to have characteristic roots which had predictable properties. Charles Hermite[6] showed that the roots of the characteristic equation of a Hermitian matrix[7] are all real. In 1885, A. Buchheim[8] proved that the characteristic roots of a symmetric[9] matrix with real elements are all real. The fact that the characteristic roots of a real skew-symmetric matrix are pure imaginary was found by Karl Weierstrass[10] in 1870.

A related result, also subsumed under Metzler's findings of 1892, was proven by Henry Taber[11]; he showed that the sum of the characteristic roots is equal to the sum of the diagonal elements (the so-called *trace* or *spur*) of the matrix.

The Cayley-Hamilton theorem states that a matrix satisfies its own characteristic equation; i.e.,

Reprinted from *Mathematics Teacher* 57 (Dec., 1962): 657–59; with permission of the National Council of Teachers of Mathematics.

if $f(x) = 0$ is the characteristic equation of a matrix M, then $f(M) = 0$.

This result was first published by Sir Arthur Cayley in 1858[12]. His original statement of the theorem was, "The determinant, having for its matrix a given matrix less the same matrix considered as a single quantity[13] involving the matrix unity, is equal to zero." Cayley proved this theorem only for second and third order matrices, stating that he felt that further proof was unnecessary. He published the proof for the two-by-two case, which in essence stated

$$\begin{vmatrix} a - \begin{pmatrix} a\ b \\ c\ d \end{pmatrix} & b \\ c & d - \begin{pmatrix} a\ b \\ c\ d \end{pmatrix} \end{vmatrix}$$

$$= \begin{pmatrix} a\ b \\ c\ d \end{pmatrix}^2 - (a+d) \begin{pmatrix} a\ b \\ c\ d \end{pmatrix}$$

$$+ (ad - bc) \begin{pmatrix} a\ b \\ c\ d \end{pmatrix}^0$$

$$= \begin{pmatrix} a^2 + bc & ab + bd \\ ac + cd & d^2 + bc \end{pmatrix}$$

$$- (a+d) \begin{pmatrix} a\ b \\ c\ d \end{pmatrix} + (ad - bc) \begin{pmatrix} 1\ 0 \\ 0\ 1 \end{pmatrix}$$

$$= \begin{pmatrix} 0\ 0 \\ 0\ 0 \end{pmatrix}.$$

To reduce the theorem to a compact form, Cayley symbolized a matrix A, considered as a single quantity, by the notation $\bar{A}$, and wrote the theorem

$$\text{Det} \ (\bar{1} \cdot M - \bar{M} \cdot 1) = 0.$$

The first general proof was given by A. Buchheim[14] in the 1884 volume of the *Messenger of Mathematics*. Sixty-three pages later in the same journal, A. R. Forsyth published a proof of the theorem for the three-by-three case.[15] Hamilton's name is connected with the Cayley-Hamilton theorem because he established the theorem for quaternions in 1853.[16]

Closely related to the characteristic polynomial of a matrix is the minimal polynomial of the matrix. The concept was put forth by Georg Frobenius[17] in 1878. His definition calls the minimal polynomial of a matrix the polynomial of least degree which the matrix satisfies. He stated that it is formed from the factors of the characteristic polynomial and that it is unique.

Seven years later, a pair of articles[18] in the *Comptes Rendus Hebdomadaires des Séances des Académie des Sciences* by Edouard Weyr discussed the problems encountered in trying to find a minimal polynomial. However, the author was unable to shed any light on a method of finding the polynomial. He coined the word "dérogatoire" to describe a matrix whose minimal polynomial has a degree which is less than the degree of the characteristic polynomial.

The first American article on the subject[19] stated that, in general, the degree of the minimal polynomial is not less than the degree of the characteristic polynomial. The author, William H. Metzler, wrote that when this happens the characteristic polynomial is said to "degrade" to a polynomial of lower degree.

In 1904, the German mathematician Kurt Hensel[20] proved that this polynomial is unique and that it divides any other polynomial which the matrix satisfies.

NOTES

1. Sir Arthur Cayley, "A Memoir on the Theory of Matrices," *London Philosophical Transactions*, CXLVIII (1858), 17–37. Also *Collected Works*, Vol. II, 475–96.

2. Mentioned by E. T. Browne in "On the Separation Property of the Roots of the Secular Equation," *American Journal of Mathematics*, LII (1930), 843–50.

3. W. H. Metzler, "On the Roots of Matrices," *American Journal of Mathematics*, XIV (1892), 326–77.

4. Kurt Hensel, "Über den Zusammenhang zwischen den Systemen und ihren Determinenten," Crelle's *Journal für reine und angewandte Mathematik*, CLIX (1928), 246–54.

5. Georg Frobenius, "Über lineare Substitutionen und bilineare Formen," Crelle's *Journal für reine und angewandte Mathematik*, LXXXIV (1878), 1–63.

6. Charles Hermite, "Remarque sur un théorème de M. Cauchy," *Comptes Rendu Hebdomadaires des Séances des Académie des Sciences*, XLI (1855), 181.

7. In a Hermitian matrix, the element a_{ij} is the complex conjugate of a_{ji}.

8. A. Buchheim, "On a Theorem Relating to Symmetrical Determinants," *Messenger of Mathematics*, XIV (1885), 143–44.

9. In a symmetric matrix, the elements a_{ij} and a_{ji} are equal.

10. This fact is taken from T. J. I'A. Bromwich, "On the Roots of the Characteristic Equation of a Linear Substitution," *Acta Mathematica*, XXX (1906), 297–304. The actual article by Weierstrass is referred to only as being on page 134 of Vol. III of his *Gesammte Werke*.

11. Henry Taber, "On the Application to Matrices of Any Order of the Quaternion Symbols S and V," *Proceedings of the London Mathematical Society*, XXII (1891), 67–79.

12. This result is found in Sec. 21 of the article mentioned in Note 1.

13. That is, considered as a unit.

14. A. Buchheim, "Mathematical Notes," *Messenger of Mathematics*, XIII (1884), 62–66.

15. A. R. Forsyth, "Proof of a Theorem by Cayley in Regard to Matrices," *Messenger of Mathematics*, XIII (1884), 139–42.

16. W. R. Hamilton, *Lectures on Quaternions* (Dublin: 1853), p. 566.

17. "Über lineare Substitutionen und belineare Formen," Crelle's *Journal für reine und angewandte Mathematik*, LXXXIV (1878), 1–63.

18. "Sur la théorie des matrices," *Comptes Rendus*, C (1885), 787–89. "Repartition des matrices en espèces et formation de toutes les espèces," *Comptes Rendus*, C (1885), 966–69.

19. "On the Roots of Matrices," *American Journal of Mathematics*, XIV (1892), 326–77.

20. "Theorie der Korper von Matrizen," *Crelle's Journal für reine und angewandte Mathematik*, CXXVII (1904), 116–66.

11

Sylvester and Scott

PATRICIA C. KENSCHAFT
AND KAILA KATZ

*T*WO BRITISH MATHEMATICIANS, both the victims of discrimination in their native country, were instrumental in stimulating the mathematical research community in the United States. James Joseph Sylvester (1814–1897) and Charlotte Angas Scott (1858–1931) inspired and trained many younger mathematicians here, served as editors of the first continuing mathematical research journal in this country, and contributed substantial research to the early American journals. Both overcame obstacles in their careers and promoted the study of mathematics by young women at a time when it was unpopular to do so.

James Joseph Sylvester

Sylvester was born into a Jewish family in London and attended private Jewish schools until the age of fourteen. After that time he was in educational settings where he often suffered because of anti-Semitism. He attended the University of London, studying under August DeMorgan, but was expelled after five months when it was discovered that he intended to use a knife against one of his tormentors. He then attended the Royal Institution at Liverpool for two years, gaining high distinction in mathematics, though at one point he ran away when harassment by fellow students became unbearable.

Reprinted from *Mathematics Teacher* 75 (Sept., 1982): 490–94; with permission of the National Council of Teachers of Mathematics.

James Joseph Sylvester (1814–1897)

In 1831, at the age of seventeen, he matriculated at Cambridge University, but illness prevented him from finishing until 1837. He placed second in the mathematical Tripos examination, a grueling five-day examination required of those who wished to graduate with honors. Unfortunately, in order to receive a degree, a student also had to subscribe to the "Thirty-Nine Articles," religious oaths of allegiance to the Church of England. Not being a member of that church, Sylvester could neither receive his degree, go on for an M.A.

at Cambridge, nor become a member of the Cambridge faculty. It wasn't until 1872, when Cambridge rescinded these religious tests, that Sylvester was awarded his long overdue degrees. In 1837 he returned to the University of London (whose name had changed to University College) to teach science, where he became a colleague of his former teacher, DeMorgan. After two years there, he left for Ireland. In 1841, Trinity College in Dublin awarded him both the B.A. and the M.A. degrees that his own alma mater would not confer upon him.

Accepting an offer to become professor of mathematics at the University of Virginia in 1841 was a disastrous decision for Sylvester. The Richmond newspapers were against the appointment of the Jewish foreigner. They took every opportunity to tell their readers of their opposition to his appointment as well as that of another foreigner who was Catholic. Some students apparently shared that opposition. An inattentive, hostile student insulted Sylvester on more than one occasion in class, and a faculty committee to which he complained failed to support him. Never one to take abuse lightly, he resigned five months after his arrival. Unable to obtain another teaching position in the States, he returned to London in 1842. He became an actuary, tutored students (including Florence Nightingale) in mathematics in his spare time, and later studied law.

It was in 1850 that he met another lawyer, Arthur Cayley, with whom he collaborated in ground-breaking research into matrices and linear transformations. Together they developed the theory of algebraic invariants, those algebraic expressions that remain invariant or unchanged under linear transformations. Students who study matrices may encounter Sylvester's law of nullity. The nullity of a matrix is the difference between its order and its rank; it is also referred to as the dimension of the kernel. Sylvester's law says that the nullity of the product of two square matrices can never exceed the sum of the nullities of the factors and is never less than the nullity of each

factor (Baker 1904). Sylvester was the first to use several words that are now common in matrix theory, including *invariant, covariant,* and *Jacobian.* Sylvester's research earned for him election as foreign correspondent to the highly prestigious and selective French Academy of Science in 1863.

In 1856 Sylvester became professor of mathematics at the Royal Academy in Woolwich, England. This position ended in 1870 when the school forced him to retire because of his age, although he was only fifty-six years old. The school then tried, unsuccessfully, to swindle him out of part of his pension. Letters to the editor of the *London Times* and a lead article there helped him secure his full pension.

Fortunately, these bitter experiences did not dull his enthusiasm for teaching. His happiest years as a teacher began in September 1876, at the age of sixty-two. The trustees of the Johns Hopkins University, seeking an internationally known mathematician for their new university, offered him a chair, which he accepted. Sylvester's successor in that chair of mathematics at Johns Hopkins described Sylvester as one who was extremely appreciative of the work of others, who was stimulated by his students' questions, and who gave the warmest recognition to any talent or ability displayed by his students.

In 1878 while at Johns Hopkins, he founded the *American Journal of Mathematics* and served as editor for its first five years. During its first ten years he contributed thirty papers to this journal. Although he spent only seven years at Johns Hopkins, he is often credited with being the individual who was most significant in the beginning of mathematical research in the United States. In particular, Thomas Fiske (1905), the founder and president of the American Mathematical Society, said in 1904 in a speech honoring the Society's tenth anniversary, "With the arrival of Professor Sylvester at Baltimore and the establishment of the *American Journal of Mathematics* began the systematic encouragement of mathematical research in America."

In 1883 Sylvester left the United States to accept an invitation from Oxford University to become the Savilian Professor of geometry. He worked there creatively until ill health forced him to retire in 1893, though he continued his mathematical research until his death in 1897.

Charlotte Angas Scott

Charlotte Scott was born in 1858. When she was a teenager, secondary education for women in England was virtually nonexistent. However, her father was president of a college and evidently provided fine tutors for her. On the basis of home study she won a scholarship in 1876 to the recently founded Girton College at Cambridge University, the first college in England open to women (Girton College Records 1876). In 1880 she obtained special permission to take the Tripos examinations with the Cambridge men, and she placed eighth in mathematics in the entire university. Because she was a woman, she could not be present at the award ceremony or even have her name read; but when the name of a young man was read in place of hers, the indignant male students shouted throughout the hall, "Scott of Girton! Scott of

Charlotte Angas Scott (1858–1931)

Girton!" The resulting publicity enabled women to gain the right to take the Tripos exams after an appropriate time of residence at the University and to have their names posted with the men.

Cambridge did not award degrees to women, however, until 1948; so Scott had to retake all her examinations for the University of London. She was awarded a bachelor's degree from that institution in 1882 and a doctorate in 1885. She had no thesis advisor in the modern sense, but evidently Arthur Cayley served in that role. In his obituary she reported that she had attended his lectures at Cambridge for four years beginning in 1880 and that ". . . for the last fourteen years I have been priviledged to know him and experience his kindness" (Scott 1895). He wrote a letter commending her mathematical research, which helped her to obtain a position on the original faculty of Bryn Mawr College immediately after receiving her doctorate (Bryn Mawr College Executive Committee Report 1884).

Thus, Scott too joined the initial faculty of a leading American institution. Her tenure was longer than Sylvester's; after serving as department head for thirty-nine years, she remained on campus without teaching classes an additional year while the last of her seven successful doctoral students completed her degree (Lehr 1974). In 1907, four of the fifteen women with doctorates on the membership list of the American Mathematical Society (AMS) had been Scott's students and a fifth was Scott herself. Many of her students who did not earn doctorates nevertheless attained important positions in higher education in our country.

Scott was a coeditor from 1899 until 1926 of the journal that Sylvester had founded. She had about forty mathematical publications, three of which appeared in that journal before she became coeditor. Although a British citizen until her death, she was a member of the original council of the AMS when it was organized in 1894. After serving for five years as a council member, she became vice-president of the AMS in 1905, and she was the only woman during the first eighty years of the AMS to serve in that position. In his semicentennial celebration speech in 1934, Thomas Fiske, reviewing the first fifty years of the society he had founded, mentioned about thirty men and only one woman, Charlotte Scott (*American Mathematical Society Semicentennial Celebration* 1939).

Scott (1894) wrote a graduate textbook in analytic geometry that was so widely used that it was reprinted by a new publisher in 1924, thirty years after its original appearance (see Bibliography). Through it alone she made a significant impact on mathematical research in our country.

Her own research involved geometrical invariants, quantities that do not change when a figure is modified or the viewpoint is altered. She was especially interested in algebraic figures in two dimensions whose equations were of degree three, four, or five (in both variables); she investigated nodes, cusps, double tangents, and "inflexions" of such curves. One of her most widely quoted papers was a geometric proof of Max Noether's theorem about quantities of the form $Af + Bg = 0$. If $f = 0$ and $g = 0$ are algebraic curves that intersect, then any curve defined by $Af + Bg$ passes through the points of intersection of f and g. Scott's (1899) geometrical proof of the converse (that all curves passing through all points of intersection of f and g must be of the form $Af + Bg$) was published in the *Mathematische Annalen*. It may have been the first mathematical paper written in the United States that received widespread attention in Europe (see Abyankar 1981).

Scott's frequent summer trips across the Atlantic Ocean at a time when such travel was difficult and rare did much to stimulate mathematics in our developing country by maintaining contact with the established world of culture. In 1922 when about 200 people gathered to honor her at Bryn Mawr College, Alfred North Whitehead made a special trip to the United States to attend. His tribute included the following sentences. "A friendship of people is the outcome of personal relations. A life's work such as that of Professor Charlotte Angas Scott is worth more to the world than many anxious efforts of diplomats. She is an example of the universal brotherhood of civilization" (Putnam 1922).

Sylvester and Scott may have been acquainted since they both worked closely with Cayley. Certainly both believed in the importance of women studying mathematics. Sylvester's tutoring of Florence Nightingale during his years as an actuary gives testimony to his belief. Later when a woman named Christine Ladd (later Ladd-Franklin) applied to become a graduate student at the Johns Hopkins University and the trustees debated whether or not to accept a woman, it was Sylvester who insisted that sex should not be a factor in judging mathematical excellence. The parallels in the lives of Sylvester and Scott and in the barriers that they had to overcome to exercise their enthusiasm for teaching and research are striking. The fact that the birth of research mathematics in this country can be largely traced to these two English mathematicians is an important message that bears repeating even now.

BIBLIOGRAPHY

Abyankar, Shreeram S. Telephone conversation on 14 February 1981, with Patricia C. Kenschaft.

American Mathematical Society Annual Register, 1907.

Baker, H. F., ed. *The Collected Mathematical Papers of James Joseph Sylvester*, IV:53. Cambridge:1904–1912, 558.

Ball, W. W. R. *A History of the Study of Mathematics at Cambridge*. Cambridge: At the University Press, 1889.

"The Beginnings of the American Mathematical Society: Reminiscences of Thomas Scott Fiske." *American Mathematical Society Semicentennial Celebration*, 1939.

Bryn Mawr College Executive Committee Report. 7 July 1884. Supplied by Lucy Fisher West, archives librarian.

Cambridge University Commission. *Report of Her Majesty's Commissioners Appointed to Inquire into the*

State, Discipline, Studies, and Revenues of the University and Colleges of Cambridge: Together with the Evidence, and an Appendix. London, Her Majesty's Stationery Office, 1852.

Cartwright, Dame Mary. Interview with Patricia C. Kenschaft, July 1974.

Catalogue of Scientific Papers, Vol. 5. London: Royal Society of London, 1871, 901–4.

Fisher, Charles S. "Death of a Mathematical Theory: A Study in the Sociology of Knowledge." Archive for History of Exact Sciences 3(1)(1966): 137–59.

Fiske, Thomas S. "Mathematical Progress in America." Bulletin of the AMS 11 (February 1905): 238–46.

Girton College Records. 1876. Supplied by Margaret Gaskell, archives librarian.

Glaisher, J. W. L. "Presidential Address; on the Mathematical Tripos." Proceedings of the London Mathematical Society 18 (11 November 1886): 4–38.

Hawkins, Hugh. Pioneer: A History of the Johns Hopkins University, 1874–1889. Cornell University Press, 1960.

Jones, E. E. Constance. Girton College. Adam and Charles Block, 1913.

Kenschaft, P. C. "Charlotte Angas Scott, 1858–1931."

Association for Women in Mathematics Newsletter 8 (April 1978): 11–12.

Kline, Morris. Mathematical Thought from Ancient to Modern Times. New York: Oxford University Press, 1972, Chapter 33.

Lehr, Marguerite. Interview with P. C. Kenschaft, June 1974.

North, John D. "Sylvester, James Joseph," Dictionary of Scientific Biographies 13 (1976): 216–22.

Putnam, Emily J. "Celebration in Honor of Professor Scott." Bryn Mawr Bulletin 2 (1922):12–14.

Scott, C. A. An Introductory Account of Certain Modern Ideas and Methods in Plane Analytical Geometry. London and New York: Macmillan and Co., 1894.

———. "Arthur Cayley, Obituary." Bulletin of the AMS 1 (March 1895): 133–41.

———. "A Proof of Noether's Fundamental Theorem." Mathematische Annalen 52 (1899): 592–97.

———. Projective Methods in Plane Analytic Geometry. New York and London: Chelsea Publishing Co., 1924.

Yates, R. C. "Sylvester at the University of Virginia." American Mathematical Monthly 44 (April 1937): 194–201.

12

The Early Beginnings of Set Theory

PHILLIP E. JOHNSON

GEORG CANTOR CREATED and largely developed the theory of sets in approximately the years 1874–1897. In contrast to such developments as the calculus and non-Euclidean geometry, the creation of set theory was, according to all indications, Cantor's alone. Also, set theory was not preceded by a long evolutionary period such as is usually the case with big mathematical breakthroughs. The present article will concern itself primarily with the very earliest set-theoretic works of Cantor, namely, his first two papers in this area.[1]

In view of Cantor's outstanding contributions to mathematics, it is surprising that he did not start out originally to become a mathematician. Cantor's father, a practical-minded, prosperous businessman, wanted Georg to go into the trade or profession of engineering so he could make a living. Georg submitted at first to his father's wishes, but his heart was never in it. Luckily for the mathematical world, his father eventually saw the folly of trying to keep such a brilliant mathematical mind as Georg's tied down to something so mundane as engineering. It may be that Georg would have eventually studied mathematics even without

his father's blessing, but certainly his father's consent made things easier for him.[2]

Cantor studied at the University of Berlin under three of the truly great mathematicians of that time: Ernst Eduard Kummer, Karl W. T. Weierstrass, and Leopold Kronecker. He received his Ph.D. degree from Berlin in 1867. His doctoral dissertation, and indeed all his works until the early 1870s, although excellent, gave no hint of the outstanding mathematical originator that he was to become.

The birth of set theory can now be recognized in an 1874 article by Cantor which appeared in *Crelle's Journal*.[3] In this paper he proved the now well known theorem that the set of real algebraic numbers can be put in one-to-one correspondence with the set of natural numbers (positive whole numbers), whereas the same is not true of the set of real numbers.[4] Up to this time different orders of infinity had not been recognized. According to Fraenkel, Cantor himself had first thought that the continuum could be put in one-to-one correspondence with the set of natural numbers.[5] Cantor remarked in the introduction to his article that by combining the two above-mentioned theorems there results a proof of the theorem first proved by Liouville that in each given interval there exist infinitely many transcendental (nonalgebraic) real numbers.[6]

Cantor's next article[7] brought forth considerable opposition, expecially from his former teacher Kronecker. A number of important theorems

Reprinted from *Mathematics Teacher* 63 (Dec., 1970): 690–92; with permission of the National Council of Teachers of Mathematics. This article is adapted from the author's Ph.D. dissertation, "A History of Cantorian Set Theory," George Peabody College for Teachers, 1968, under the direction of Dr. J. Houston Banks.

concerning equivalent sets appeared in this paper.[8] One of the very well known results proved here is that the set of rational numbers can be put in one-to-one correspondence with the set of natural numbers.[9] Another interesting result (even though now obvious) noted by Cantor is something that had been noticed by Bernard Bolzano several years earlier: For infinite sets a set may be equivalent to a (proper) subset of itself; for example, the set of natural numbers is equivalent to the set of even natural numbers.[10, 11] Cantor also conjectured in this paper that the two powers of the rational numbers and the real numbers exhaust all possibilities for infinite subsets of the continuum.[12] Time has shown that he was overly optimistic, in that it has now been shown in the realm of axiomatic set theory that the conjecture is neither provable nor disprovable. Perhaps the result most objected to was Cantor's proof of the independence of the power of the continuum from its number of dimensions. That this result surprised even himself is evident in one of his letters to his close friend Richard Dedekind prior to publication of the proof.[13] It had been commonly assumed that points in two-space cannot be traced back to one-space, yet Cantor's proof said that the set of points in two-space is equivalent to the set of points in one-space. In fact, n-space is equivalent to one-space, and the result can even be expanded to the case of a countable infinity of dimensions.[14] These are some of the results in Cantor's second paper on set theory.

Cantor's second paper on set theory had rough sledding at *Crelle's Journal*, and he never published another paper in that journal. For a while it appeared that the paper would not be published when it was slated to be, apparently because of Kronecker's rejecting the point of view of Cantor's ideas (Kronecker was on the editing staff of the *Journal*).

Following Cantor's two papers that represent the earliest beginnings of set theory was an intensive working period by him in the years 1879–1884, during which he published practically his complete theory of sets. Cantor faced tremendous opposition in gaining recognition of these works. Kronecker and a number of other prominent mathematicians of that time were firmly aligned against his new and strange notions.

Cantor suffered a complete breakdown in the spring of 1884. In retrospect a number of contributing causes can be seen for this, among which were the trouble in getting his important paper of 1878 accepted, the hard struggle to gain recognition for his works of 1879–1884, and the formidable array of influential colleagues against his works. In addition, he was only moderately satisfied with the position that he occupied at Halle and would have preferred the wider field of work offered by the University of Berlin. As long as Kronecker was at Berlin, however, there was little chance that Cantor could get an appointment there.

Although Cantor's mental illness recurred throughout his life, his mental crisis was essentially over at the beginning of 1885; and his confidence in his work, which had been shaken, was reestablished. He published papers in set theory until 1897. Some of this work was particularly noteworthy in crystallizing some of his previous notions.

Hardly had Cantor's work been completed before paradoxes began to appear.[15] Despite the paradoxes and the difficulties in gaining recognition of his work, the creation of set theory is an undeniably important mathematical development. Fortunately Cantor lived long enough to see the beginning of the tremendous impact which his theory was destined to have on the mathematical world and to enjoy the pleasure of the belated recognition that he so much deserved.

NOTES

1. There was one earlier paper that contains some of the rudimentary concepts out of which Cantor's consideration of general point sets grew, but this is not being considered as his first paper on set theory since the paper was really on trigonometric series. The earlier

paper was "Ueber die Ausdehnung eines Satzes aus der Theorie der trigonometrischen Reihen," *Mathematische Annalen* 5 (1872): 123–32.

2. The biographical information about Cantor in this article is chiefly from Abraham Adolf Fraenkel's excellent biography, "Georg Cantor," *Jahresbericht der Deutschen Mathematiker Vereinigung* 39 (1930): 189–266.

3. Georg Cantor, "Ueber eine Eigenschaft des Inbegriffes aller reellen algebraischen Zahlen," *Journal für die reine und angewandte Mathematik* 77 (1874): 258–62.

4. The proofs of these two theorems are readily available in a number of sources and will not be repeated here. The proof (using Cantor's famous diagonal process) usually given of the second theorem mentioned is not the one that Cantor gave at this time but is due to a later work.

5. Fraenkel, "Georg Cantor," p. 237.

6. Ibid., p. 259.

7. Georg Cantor, "Ein Beitrag zur Mannigfaltigkeitslehre," *Journal für die reine und angewandte Mathematik* 84 (1878): 242–58.

8. Two sets are *equivalent* if their elements can be placed in one-to-one correspondence with each other.

9. The proof of this result also is readily available and will not be repeated.

10. Georg Cantor, *Contributions to the Founding of the Theory of Transfinite Numbers*, a translation by Philip E. B. Jourdain of two of Cantor's works published in 1895 and 1897 (New York: Dover Publications, n.d.), p. 41. The book is also provided with an introduction and notes by Jourdain. The introduction deals in part with the work of Cantor from 1870 to 1895 and is an especially helpful source for the material in this article.

11. Cantor used "subset" in the sense that "proper subset" is now used.

12. The *power (cardinal number)* of a set is the collection of all sets that are equivalent to it.

13. Fraenkel, "Georg Cantor," p. 237.

14. Cantor did not use the term "countable" until later. A set is *countable (denumerable)* if it is equivalent to the set of natural numbers.

15. The first published paradox of set theory was the Burali–Forti paradox of 1897, and in the next few years paradoxes began to appear in abundance. The paradoxes have been of considerable importance in motivating study in the foundations of mathematics and in the axiomatizations of set theory by such people as Ernst Zermelo, Fraenkel, John von Neumann, Paul Bernays, and Kurt Gödel.

13

Infinity: The Twilight Zone of Mathematics

WILLIAM P. LOVE

*T*HE CONCEPT OF INFINITY has fascinated the human race for thousands of years. Who among us has never been awed by the mysterious and often paradoxical nature of the infinite? The ancient Greeks were fascinated by infinity, and they struggled with its nature. They left for us many unanswered questions including Zeno's famous paradoxes. The concept of infinity is with us today, and many ideas in modern mathematics are dependent on the infinitely large or the infinitely small. But most people's ideas about infinity are very vague and unclear, existing in that fuzzy realm of the twilight zone.

Our mathematics curriculum includes the study of infinite sets of numbers in algebra and infinite sets of points in geometry, but students rarely understand how these may be related.

Ask your students these questions:

1. Which is larger: the set of all prime numbers or the set of all composite numbers?
2. Which is larger: the set of all rational numbers or the set of all irrational numbers?
3. Which is larger: the number of points in the interval [0, 1] or the number of points in the interval [0, 2]?
4. Which is larger: the number of points in a line or the number of points in a plane?

Reprinted from *Mathematics Teacher* 82 (Apr., 1989): 284–92; with permission of the National Council of Teachers of Mathematics.

Georg Cantor

The photograph of Georg Cantor appears through the courtesy of Ivor Grattan-Guinness.

Most students will answer these questions incorrectly because their intuitive understanding of finite sets does not always apply to infinite sets. In addition, most students are unaware that there are infinitely many different infinities.

Unless some ideas involving mathematical infinities are included in their instruction, students will never develop a true understanding of the various number systems introduced in their algebra classes nor will they understand how these sets relate to points, lines, and planes introduced in their geometry classes.

This article presents in a clear and understandable manner the main ideas concerning the theory

THE SEARCH FOR CERTAINTY

of infinity as developed by Cantor and others. It includes most of the important theorems and a sketch of their proofs. Details of proofs may be found in advanced textbooks on set theory.

This material could be used as a reading assignment for interested students, as a topic for presentation at a mathematics club meeting, or as a source of information for teachers. Students might find it an interesting topic for a mathematics fair project. A teacher might challenge students to find their own proofs for some of the easier theorems presented here.

Let us examine infinity more carefully.

Georg Cantor (1845–1918) was one of the first modern mathematicians to seriously examine the infinite. His work was not accepted by everyone, and many considered him insane. He received little recognition in his lifetime and eventually died in an asylum. Today he is considered to be one of the founding fathers of the theories of the infinite, but even now there are some who do not accept his work. He began by looking at properties of sets, especially the cardinality of sets.

Cardinality of Sets

DEFINITION. *The cardinality of a set S is the number of elements in that set. This is denoted by n(S). For example, if F = {fingers on a hand}, then n(F) = 5.*

Sets A and B have the same cardinality if $n(A) = n(B)$. For small finite sets, it is easy to determine if two sets have the same number of elements by simply counting them. But when the sets are large it is not so easy. For example, in a large auditorium, are there more seats or more people? Rather than count them, it is easier to pair each person with one seat and then see if any people or seats are left over. This principle applies to both finite and infinite sets.

DEFINITION. *A one-to-one correspondence between sets A and B is a pairing of the elements of one set with the elements of the other set in such a way that all elements from both sets have exactly one partner.*

This is denoted by $A \leftrightarrow B$.

Thus, if it is possible to find a one-to-one correspondence between set A and set B, then A and B have the same number of elements and have the same cardinality; that is,

$$\text{if } A \leftrightarrow B, \text{ then } n(A) = n(B).$$

In addition,

$$\text{if } A \text{ is a subset of } B, \text{ the } n(A) \le n(B).$$

The First Transfinite Number: Aleph Null

In the 1880s, Cantor applied these ideas to infinite sets. He began by examining the infinite set of natural numbers, denoted by N. He defined the cardinality of the set of natural numbers to be *aleph null*, denoted $\aleph_0$, using the first letter of the Hebrew alphabet, $\aleph$, followed by the subscript zero. Thus, $n(N) = \aleph_0$. This is called a *transfinite* number, being "beyond the finite."

Cantor began to examine various sets of numbers to determine their cardinality relative to the natural numbers. He reasoned that if it was possible to find a one-to-one correspondence between a set S and the set of natural numbers, then they must have the same number of elements; that is,

$$\text{if } S \leftrightarrow N, \text{ the } n(S) = \aleph_0.$$

THEOREM 1. *Cardinality of the set of all even numbers is $\aleph_0$.*

Proof. If evens $\leftrightarrow$ naturals, then $n(\text{even}) = \aleph_0$.

even:	2	4	6	8	10	...	2k	...
	↓	↓	↓	↓	↓		↓	
natural:	1	2	3	4	5	...	k	...

Therefore,

$$n(\text{even}) = \aleph_0.$$

THEOREM 2. *Cardinality of all nonnegative integral powers of 10 is $\aleph_0$.*

Proof. If powers of 10 $\leftrightarrow$ naturals, then n(powers of 10) = $\aleph_0$.

10^p:	10^0	10^1	10^2	10^3	10^4	...	10^k	...
	$\downarrow$	$\downarrow$	$\downarrow$	$\downarrow$	$\downarrow$		$\downarrow$	
natural:	1	2	3	4	5	...	$k+1$	...

Therefore,
$$n(\text{powers of }10) = \aleph_0.$$

THEOREM 3. *Cardinality of all prime numbers is $\aleph_0$.*

Proof. If primes $\leftrightarrow$ naturals, then n(primes) = $\aleph_0$.

primes:	2	3	5	7	11	13	...	P_k	...
	$\downarrow$	$\downarrow$	$\downarrow$	$\downarrow$	$\downarrow$	$\downarrow$		$\downarrow$	
natural:	1	2	3	4	5	6	...	k	...

Although there is no formula for the pairing, the kth prime is paired with the natural number, k. Therefore,

$$n(\text{primes}) = \aleph_0.$$

THEOREM 4. *Every infinite subset of the natural numbers has cardinality $\aleph_0$.*

The student should recognize the implications of theorem 4. This means that all the following sets have the same number of elements: natural numbers, even numbers, odd numbers, prime numbers, composite numbers, multiples of 5, multiples of 10, perfect squares, perfect cubes, triangular numbers, Fibonacci numbers, and any other infinite subset of the naturals. All these have cardinality $\aleph_0$.

This implication violates one's intuition. How can a proper subset have the same number of elements as the entire set? One of the basic assumptions taught since the time of the ancient Greeks was that "the whole is greater than the

part." This notion is true for finite sets but not true for infinite sets. In 1888, Dedekind proposed this assumption as the basis for a definition of an infinite set.

DEFINITION. *A set S is infinite if and only if S has the same number of elements as one of its proper subsets.*

Cantor next examined sets that were "larger" than the set of natural numbers.

THEOREM 5. *Cardinality of the set of integers is $\aleph_0$.*

Proof. In integers $\leftrightarrow$ naturals, then n(integers) = $\aleph_0$. The positive integers are paired with the even naturals, and the nonpositive integers are paired with the odd naturals.

integers:	...	-2	-1	0	1	2	...
		$\downarrow$	$\downarrow$	$\downarrow$	$\downarrow$	$\downarrow$	
naturals:	...	5	3	1	2	4	...

That is, whenever $k > 0$, $k \rightarrow 2k$; and when $k \leq 0$, $k \rightarrow -2k + 1$. Therefore,

$$n(\text{integers}) = \aleph_0.$$

Next, Cantor investigated the set of rational numbers. The rationals have the density property so that between any two distinct rationals there is an infinite number of other rationals. Students' intuition will tell them that there are more rational numbers than natural numbers, but Cantor proved that this assumption is false by using an unusual "diagonal process."

THEOREM 6. *Cardinality of the set of rational numbers is $\aleph_0$.*

Proof. If rationals $\leftrightarrow$ naturals, then n(rationals) = $\aleph_0$. Cantor used arrays to list all possible forms a/b. The positive array (see Fig. 1) includes all positive rational numbers, and a similar nonpositive array may be constructed to include all other ratio-

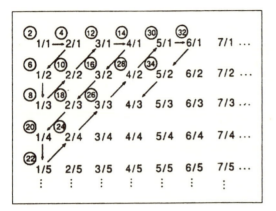

FIGURE I.

Cantor's one-to-one correspondence between the positive rational numbers and the even natural numbers

nal numbers. Cantor established a one-to-one correspondence between the positive array and even numbers as shown in Figure 1. Thus, positive array ↔ evens. In a similar manner, it is possible to construct the nonpositive array such that nonpositive array ↔ odds. Thus, combined arrays ↔ naturals, so n(combined arrays) = $\aleph_0$.

Since 1/2, 2/4, 3/6, . . . all represent the same rational number, we see that the combined arrays contain the set of all rational numbers. Thus,

$$n(\text{rationals}) \leq n(\text{combined arrays}) = \aleph_0.$$

The natural numbers are a subset of the rationals. Therefore,

$$\aleph_0 = n(\text{naturals}) \leq n(\text{rationals}).$$

Thus we have that

$$n(\text{rationals}) \leq \aleph_0,$$

and that

$$\aleph_0 \leq n(\text{rationals}).$$

Therefore,

$$n(\text{rationals}) = \aleph_0.$$

THEOREM 7. *If S is any infinite set whose elements may be uniquely determined by an ordered pair of integers $\{a, b\}$, then cardinality of S is $\aleph_0$.*

The proof is similar to that for theorem 6, since $\{a, b\}$ may replace a/b.

THEOREM 8. *If S is any infinite set whose elements may be uniquely determined by a finite ordered n-tuple $\{a_1, a_2, a_3, ..., a_n\}$, then cardinality of S is $\aleph_0$.*

The proof is left to the reader.

Next, Cantor investigated the set of irrational numbers. The irrational numbers consist of algebraic numbers (such as $\sqrt{2}$, $\sqrt{3}$, $\sqrt[3]{5}$) and transcendental numbers (such as π and e).

DEFINITION. *A number, k, is algebraic if and only if k is the solution to some algebraic equation of the form $a_n x^n + a_{n-1} x^{n-1} + ... + a_1 x + a_0 = 0$, where all a_i are integers.*

For example, the algebraic equation $x^2 - 2 = 0$ has two solutions: $x_1 = \sqrt{2}$ and $x_2 = -\sqrt{2}$. Thus, both are algebraic numbers. The first root, x_1, may be characterized by the set of coefficients of the equation $(1, 0, -2)$ plus the root number = 1:

$$x_1 \leftrightarrow \{1, 0, -2, 1\}$$

The second root, x_2, may be characterized by the set of coefficients of the equation $(1, 0, -2)$ plus the root number = 2:

$$x_2 \leftrightarrow \{1, 0, -2, 2\}$$

THEOREM 9. *Cardinality of the set of algebraic numbers is $\aleph_0$.*

Proof. Every algebraic number is the solution to some algebraic equation having a finite number of coefficients. Thus, each algebraic number can be uniquely characterized by a finite ordered n-tuple consisting of coefficients plus root number.

Therefore, by theorem 8, the cardinality of the set of all algebraic numbers = $\aleph_0$. An interesting result is observed by using some examples: n(even naturals) = $\aleph_0$; n(odd naturals) = $\aleph_0$; and

n(even naturals $\cup$ odd naturals)

$$= n(\text{naturals}) = \aleph_0.$$

This implies that

$$\aleph_0 = \aleph_0 = \aleph_0.$$

In general, any finite number of $\aleph_0$'s added together results in $\aleph_0$.

Cantor wondered if all infinite sets had cardinality $\aleph_0$. He spent three years trying to show that the cardinality of points on a line was $\aleph_0$, but he never succeeded. Instead, he made the most surprising discovery of his life. He found a new infinity.

A Second Transfinite Number: c

THEOREM 10. *Cardinality of points in interval* $(0, 1)$ *is not* $\aleph_0$.

Proof. If $(0, 1) \leftrightarrow N$ is impossible, then $n(0, 1)$ $\neq \aleph_0$. To show this impossibility, Cantor had to

TABLE I

Cantor's Typical Correspondence between Natural Numbers and Decimal Number in (0, 1)

Natural Numbers	Decimal Number in (0, 1)
1	$\leftrightarrow$ 0. $a_1 a_2 a_3 a_4 a_5 a_6 \ldots$
2	$\leftrightarrow$ 0. $b_1 b_2 b_3 b_4 b_5 b_6 \ldots$
3	$\leftrightarrow$ 0. $c_1 c_2 c_3 c_4 c_5 c_6 \ldots$
$\vdots$	$\vdots$
k	$\leftrightarrow$ 0. $k_1 k_2 k_3 k_4 k_5 k_6 \ldots$
$\vdots$	$\vdots$

show that every possible correspondence $(0, 1) \leftrightarrow N$ would not work. To do this, he let each real number in $(0, 1)$ be written as an infinite decimal (such as $0.38 = 0.379\,999\,99 \ldots$). Then every such pairing could be shown as in Table 1. For every possible correspondence between N and $(0, 1)$ one can construct a decimal number z in the interval $(0, 1)$ that is not in the table and is not paired with any natural number. Define

$$z = 0.z_1 z_2 z_3 z_4 z_5 \ldots$$

as follows:

$$z_1 \neq a_1$$

Hence, z is not the number paired with 1.

$$z_2 \neq b_2$$

Hence, z is not the number paired with 2.

$$z_3 \neq c_3$$

Hence, z is not the number paired with 3. And so on. Thus, z is not paired with any natural number in N, so the correspondence $(0, 1) \leftrightarrow N$ is impossible. Therefore, $n(0, 1) \neq \aleph_0$.

Since $n(0, 1)$ is obviously infinite and $n(0, 1) \neq \aleph_0$, there must exist a transfinite cardinal number different from $\aleph_0$ — a second infinity. This discovery shocked the mathematical world. Cantor named the new transfinite number c for the "continuum."

DEFINITION. *Cardinality of points in interval* $(0, 1)$ *is c.*

With Cantor's discovery, two transfinite numbers had been identified: $\aleph_0$ and c. In addition, $\aleph_0$ is the "smallest" infinity and $\aleph_0 < c$. Any set having cardinality $\aleph_0$ is said to be *countably infinite*, and any set having cardinality c is *uncountably infinite*.

THEOREM 11. *Cardinality of points on interval* [0. 1] *is c.*

Proof. If $[0,1] \leftrightarrow (0, 1)$, then $n[0,1] = c$. This correspondence is established by three 1–1 onto functions:

$$
\begin{array}{cccc}
 & f & g & h \\
[0,1] & \leftrightarrow [0,1) & \leftrightarrow (0, 1] & \leftrightarrow (0, 1)
\end{array}
$$

1. First, consider the function h:

Define $h_1(x) = 3/2 - x$ where $x \in (1/2, 1]$, $y \in$ [1/2, 1).

Define $h_2(x) = 3/4 - x$ where $x \in (1/4, 1/2\}$, $y \in [1/4, 1/2)$.

Define $h_3(x) = 3/8 - x$ where $x \in (1/8, 1/4]$, $y \in [1/8, 1/4)$.

Define $h_i(x) = 3/2^i - x$ where $x \in (1/2^i, 1/2^{i-1}]$, $y \in [1/2^i, 1/2^{i-1})$.

We see that $h(x) = h_1(x) \cup h_2(x) \cup h_3(x) \cup \ldots \cup h_i(x)$ … defines a one-to-one correspondence $(0, 1] \leftrightarrow (0, 1)$.

2. Second, consider the function g:

Define function $g(x) = 1 - x$. Clearly, $g(x)$ defines a one-to-one correspondence $[0, 1) \leftrightarrow (0, 1]$.

3. Third, consider the function f:

Define the function $f(x)$ such that $f(0) = 0$ and $f(x) = h(x)$ for $x \in (0, 1]$ as defined in part 1 of this proof. Clearly, f defines a one-to-one correspondence $[0, 1] \leftrightarrow [0, 1)$.

4. Thus, we have

$$[0, 1] \leftrightarrow [0, 1) \leftrightarrow (0, 1] \leftrightarrow (0, 1).$$

Hence, $[0, 1] \leftrightarrow (0, 1)$. Therefore, $n[0, 1] = c$.

Next, Cantor showed that all segments have the same number of points.

THEOREM 12. *Cardinality of points on interval* $[a, b]$ *is c.*

Proof. If $[a, b] \leftrightarrow [0, 1]$, then $n[a, b] = c$.

FIGURE 2

Geometric correspondence between points on [0, 1] and [a, b]

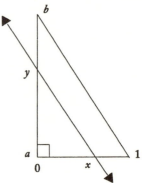

Geometric proof. Construct a right triangle from segments $[0, 1]$ and $[a, b]$ with 0 and a at the right angle. Construct a hypotenuse through 1 and b. Every line parallel to the hypotenuse that intersects the triangle will pair a point x in $[0, 1]$ with exactly one point y in $[a, b]$ as shown in Figure 2. Hence, $[a, b] \leftrightarrow [0, 1]$. Therefore $n[a, b] = c$.

Algebraic proof. Define $f(x) = (x - a)/(b - a)$, where $x \in [a, b]$ and $f(x) \in [0, 1]$. So f defines a one-to-one correspondence $[a, b] \leftrightarrow [0, 1]$. Therefore, $n[a, b] = c$.

COROLLARY. *All segments have the same number of points.*

Next, Cantor showed that lines and segments have the same number of points.

THEOREM 13. *Cardinality of points on line* $(-\infty, +\infty)$ *is c.*

Proof. If $(0, 1) \leftrightarrow (-\infty, +\infty)$, then $n(-\infty, +\infty) = c$.

Geometric proof. The interval $(0, 1)$ may be shaped into a semicircle having center P and at the point 0.5 being tangent to a line. Every ray from P that intersects the semicircle will also intersect the line and thus will pair exactly one $x \in (0, 1)$ with exactly one $y \in (-\infty, +\infty)$ as shown in Figure 3.

FIGURE 3

Geometric correspondence between points on [0, 1] and $(-\infty, +\infty)$

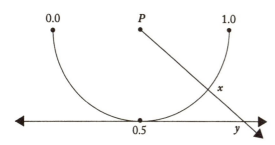

Algebraic proof. We know that

$$(0, 1) \leftrightarrow \left(-\frac{\pi}{2}, +\frac{\pi}{2}\right)$$

from theorem 12. The function $f(x) = \tan(x)$ defines a correspondence

$$\left(-\frac{\pi}{2}, +\frac{\pi}{2}\right) \leftrightarrow (-\infty, +\infty);$$

so

$$(0, 1) \leftrightarrow (-\infty, +\infty).$$

Thus,

$$n(-\infty, +\infty) = c.$$

This means n(real numbers) $= c$.

COROLLARY. *All segments, rays, lines, circles, triangles, squares, rectangles, polygons, and plane curves have the same number of points, namely c.*

COROLLARY. *Cardinality of irrational numbers is c.*

Proof. Since

$$\text{reals} = \text{rational} \cup \text{irrational},$$

we have

$$n(\text{reals}) = n(\text{rationals}) + n(\text{irrationals}).$$

Thus,

$$c = \aleph_0 + n(\text{irrationals}).$$

If n(irrationals) $= \aleph_0$, then we have $\aleph_0 + \aleph_0 = c$, which is false. Thus, n(irrationals $\neq \aleph_0$. This means that n(irrationals) $> \aleph_0$, and assuming the next largest transfinite number is c (i.e., assuming the continuum hypothesis), we conclude that n(irrationals) $= c$.

COROLLARY. Cardinality of transcendental numbers is c.

Proof. Since

$$\text{reals} = \text{algebraic} \cup \text{transcendental},$$

we have

$$n(\text{reals}) = n(\text{algebraic}) + n(\text{transcendental}).$$

Thus,

$$c = \aleph_0 + n(\text{transcendental}).$$

Using the same proof as for the previous corollary, we conclude that n(transcendental) $= c$.

Cantor proved that transcendental numbers existed and that they were more numerous than rational or algebraic numbers.

Next, Cantor examined points in the plane.

Definition. *The unit square is the set of all points in the plane defined by the Cartesian product $(0, 1) \times (0, 1)$. If $P = (x, y)$ is a point in the unit square, then its coordinates may be written in decimal form:*

$$P = (0.x_1 x_2 x_3 \ldots, 0.y_1 y_2 y_3 \ldots)$$

THEOREM 14. *Cardinality of all points in unit square is c.*

Proof. If unit square $\leftrightarrow (0, 1)$, then n(unit square) $= c$. Form a one-to-one correspondence as follows:

$$P = (0.x_1 x_2 x_3 \ldots, 0.y_1 y_2 y_3 \ldots)$$
$$\leftrightarrow n = 0.x_1 y_1 x_2 y_2 x_3 y_3 \ldots$$

For example,

$$P = (0.123\,4\ldots, 0.987\,6\ldots)$$
$$\leftrightarrow n = 0.192\,837\ldots$$

In this manner every point in the unit square is paired with exactly one number in interval $(0, 1)$. Thus, n (unit square) $= c$.

COROLLARY. *Cardinality of points in every plane region having area greater than zero is c.*

COROLLARY. *Cardinality of all points in a plane is c.*

COROLLARY. *Cardinality of all complex numbers is c.*

Proof. There exists a one-to-one correspondence between every complex number $a + bi$ and every point (a, b) in the plane. Thus, n(complex numbers) $= c$.

In a similar manner, Cantor turned his attention to sets of points in three-dimensional space.

DEFINITION. *The unit cube is the set of all points in space defined by the Cartesian product $(0, 1) \times (0, 1) \times (0, 1)$. If $P = (x, y, z)$ is a point in the unit cube, then its coordinates can be written in decimal form:*

$$P = (0.x_1 x_2 x_3 \ldots, 0.y_1 y_2 y_3 \ldots,$$
$$0.z_1 z_2 z_3 \ldots)$$

THEOREM 15. *Cardinality of points in a unit cube is c.*

Proof. If unit cube $\leftrightarrow (0, 1)$, then n(unit cube) $= c$. Form a one-to-one correspondence as follows:

$$P = (0.x_1 x_2 \ldots, 0.y_1 y_2 \ldots, 0.z_1 z_2 \ldots)$$
$$\leftrightarrow n = 0.x_1 y_1 z_1 x_2 y_2 z_2 \ldots$$

Example:

$$P = (0.123 \ldots, 0.456 \ldots, 0.789 \ldots)$$
$$\leftrightarrow n = 0.147\ 258\ 369 \ldots$$

In this manner, every point in the unit cube is paired with exactly one number in $(0, 1)$. Therefore, n(unit cube) $= c$.

COROLLARY. *Cardinality of all points in every geometric solid having volume greater than zero is c.*

COROLLARY. *Cardinality of all points in three-dimensional space is c.*

COROLLARY. *Cardinality of all points in N-dimensional space is c.*

Thus, the same number of points occur in the interval $[0, 1]$ as occur in all of space.

Cantor wondered if $\aleph_0$ and c were the only transfinite numbers, but another was to be discovered.

A Third Transfinite Number: d

Another transfinite number was found to be even larger than either $\aleph_0$ or c.

DEFINITION. Let

$$F = \{\text{all real functions } f \mid f: (0, 1) \to (0, 1)\}.$$

THEOREM 16. *Cardinality of F does not equal c.*

Proof. If $F \leftrightarrow (0, 1)$ is impossible, then $n(F) \neq c$. One must show that every 1–1 correspondence between F and $(0, 1)$ is impossible. Every such correspondence would pair each z in $(0, 1)$ with some function f_z in F as shown:

$$z \in (0, 1) \quad \leftrightarrow \quad f_z \in F$$
$$0.25 \in (0, 1) \quad \leftrightarrow \quad f_{0.25} \in F$$
$$0.38 \in (0, 1) \quad \leftrightarrow \quad f_{0.38} \in F$$
$$\ldots$$

For every possible correspondence between $(0, 1)$ and F, it is always possible to define a function $g \in F$ that is not in the table and is not paired with any real number in $(0, 1)$. For g and f_z to be different, they have to be unequal at one or more points. For example, if $g(x) \neq f_{0.25}(x)$ at $x = 0.25$ then g and $f_{0.25}$ are different. Define g as follows:

$$g(x) \neq f_z(x) \text{ at } x = z$$

for every $z \in (0, 1)$. Hence, g is different from every function in the table at one point and so is not paired with any number in $(0, 1)$.

Therefore, $F \leftrightarrow (0, 1)$ is impossible and $n(F) \neq c$.

Since F has more elements than $(0, 1)$, then $n(F) > c$ and is a larger transfinite number.

DEFINITION. *Cardinality of $F = \{f \mid f : (0, 1) \to (0, 1)\}$ is d.*

Thus, there exist at least three transfinite cardinal numbers:

$$\aleph_0 < c < d$$

COROLLARY. *Cardinality of $\{f \mid f : R \to R\}$ is d.*

In the plane exist infinitely many different circles, squares, rectangles, polygons, lines, rays, segments, curves, and figures. Infinitely many sets of points are possible in the plane.

THEOREM 17. *Cardinality of all sets of points in a plane is d.*

COROLLARY. *Cardinality of all sets of points in space is d.*

Infinitely Many Transfinite Numbers

If three transfinite numbers exist, it is a natural question to ask if more exist also. The answer is yes, more do exist.

Cantor proved a remarkable theorem comparing the cardinality of a set S and the cardinality of the set of all subsets of S. He showed that if $n(S) = A$, then $n(\text{all subsets of } S) = 2^A$ and $A < 2^A$. Therefore, if the cardinality of the natural numbers is $\aleph_0$, the cardinality of the set of all subsets of naturals must be $2^{\aleph_0}$. He called this new number *aleph 1*, $\aleph_1$. Since there exists a set having cardinality $\aleph_1$, then the cardinality of all subsets of it would be $2^{\aleph_1}$, which he called *aleph 2*, $\aleph_2$. Continuing in this way, Cantor showed that an infinite number of different transfinite numbers exists. He named them all alephs and formed the sequence

$$\aleph_0 < \aleph_1 < \aleph_2 < \aleph_3 < \aleph_4 < \cdots .$$

How do the transfinite numbers c and d fit with all the alephs? Cantor believed that $\aleph_1 = c$ and $\aleph_2 = d$,

but he was not able to prove it. This belief was called the *continuum hypothesis*. An equivalent form of the continuum hypothesis states that no transfinite numbers occur between $\aleph_0$ and c. A generalized continuum hypothesis assumes that the sequence of alephs are the only transfinite numbers.

Many mathematicians tried to prove the continuum hypothesis, and it was considered one of the most important unsolved problems in mathematics in 1900. The question of the continuum hypothesis has an interesting history and was not resolved until 1963 by Paul Cohen. It is now accepted as an axiom in modern set theory. However, by rejecting this axiom a new system can be created called *non-Cantorian set theory*.

Summary

1. All the following sets have cardinality $\aleph_0$:
   - Natural numbers, even numbers, and odd numbers
   - Multiples of 3, 4, 5, 6, and so on
   - Squares, cubes, and powers
   - Primes and composites
   - All infinite subsets of naturals
   - Integers, rational numbers, and algebraic numbers

2. All the following sets have cardinality c (or $\aleph_1$):
   - Transcendental numbers and irrational numbers
   - Real numbers and complex numbers
   - Points on all segments, rays, and straight lines
   - Points on every circle, polygon, or plane curve
   - Points on every unit square, polygonal region, or plane region
   - Points on every plane
   - Points in every sphere, cube, or geometric solid
   - Points in three-dimensional space or n-dimensional space

3. All the following sets have cardinality d (or $\aleph_2$):
   - All real-valued functions $f: (0, 1) \rightarrow (0, 1)$
   - All real-valued functions $g: R \rightarrow R$
   - All possible sets of points in the plane
   - All possible sets of points in three-dimensional space.

Conclusion

By learning about infinity, our students will understand a number of ideas fundamental to mathematics:

1. That properties that hold true for finite sets may not necessarily hold true for infinite sets

2. That for infinite sets, a set and a proper subset may have the same number of elements

3. That there is not just one infinity but rather infinitely many different infinities

4. That there is a fundamental difference between the infinite sets of the integers, natural numbers, and rational numbers as compared to the infinite sets of the irrational, real, and complex numbers.

BIBLIOGRAPHY

Abian, Alexander. *The Theory of Sets and Transfinite Arithmetic*. Philadelphia: W. B. Saunders Co., 1965.

Cantor, George. *Contributions to the Founding of the Theory of Transfinite Numbers*. New York: Dover Publications, 1915.

Cohen, Paul J., and Reuben Hersh. "Non-Cantorian Set Theory." In *Mathematics in the Modern World*. San Francisco: W. H. Freeman & Co., 1968.

Dauben, Joseph W. *Georg Cantor*. Cambridge, Mass.: Harvard University Press, 1979.

Davis, Paul J., and Reuben Hersh. *Mathematical Experience*. Boston: Birkhäuser Boston, 1981.

Drake, Frank R. *Set Theory—an Introduction to Large Cardinals*. Amsterdam: North Holland Publishing Co., 1974.

Hallett, Michael. *Cantorian Set Theory and Limitation in Size*. Oxford: Clarendon Press, 1984.

Kinsolving, May R. *Set Theory and the Number System*. Scranton, Pa.: International Textbook Co., 1967.

Lin, Shwu-Yeng, T. *Set Theory—an Intuitive Approach*. Boston: Houghton-Mifflin Co., 1974.

Maor, Eli. *To Infinity and Beyond: A Cultural History of the Infinite*. Boston: Birkhäuser Boston, 1987.

14

Irrationals or Incommensurables V: Their Admission to the Realm of Numbers

PHILLIP S. JONES

$\mathcal{W}$E HAVE SEEN in earlier notes[1] that incommensurable quantities were first discovered in geometric situations and were proved to exist by using the theory of evens and odds. As roots of equations, they were represented by symbols with which operations were performed, and they were approximated by rational numbers. Viete's and Descartes's literal symbolism helped to free these numbers from what in one sense was a too close association with geometric and physical magnitude. One the other hand Fermat's and Descartes's analytic geometry produced a need to associate a number with every point of a line—a feat impossible without the recognition of irrational numbers as abstract entities on a par with other numbers. This was a concept significantly different from the Greek idea of incommensurable magnitudes for whose ratios Eudoxus derived a treatment.

We might note parenthetically, however, that one French geometer, Legendre (1752–1833), in writing an elementary geometry to replace Euclid as a school text, proved propositions on similitude by applying algebraic-numerical reasoning to lit-

eral symbols which represented lengths. He thus turned the tables on the Greeks, who used lengths to represent numbers. But Legendre did not give a clear or modern exposition of irrationals.

A second motivation which we can now see would drive mathematicians to seek to clarify the idea of the irrational was the development of the calculus with its limits and problems of continuity. Augustin Cauchy, one of the prime movers in the rigorization of the calculus, in his *Cours d'Analyse* of 1821 treated irrationals as essential quantities which were familiar to all. He remarked that an irrational number could be the limit of a sequence of rationals. Although this is essentially the idea used later by Cantor, who *defined* irrationals as the limits of sequences of rationals, Cauchy's remark was logically incomplete since he did not give a definition of irrational numbers. Bernard Bolzano (ca. 1817) and C. F. Gauss were also thinking along this line.[2]

The persons responsible for the two chief theories of irrationals in use today are the Germans Georg Cantor (1845–1918) and J. W. R. Dedekind (1831–1916), both of whom published their first papers on this topic in 1872. Cantor's approach is somewhat similar to that propounded by the Frenchman C. Meray in 1869. His work is also

Reprinted from *Mathematics Teacher* 49 (Nov., 1956): 541–43; with permission of the National Council of Teachers of Mathematics.

similar, but less so, to the development of the German scholar, Karl Weierstrass, whose lectures of 1865–1866 on this topic were expanded and published by several of his students.

Cantor worked with sequences of rational numbers. Through definitions, he created a new number field out of the raw materials (rational numbers) which were on hand and understood. This process of extending old concepts or creating new ideas from old materials has happened repeatedly in the history of mathematics.

Cantor called a *fundamental sequence* any sequence of rational numbers $(a_1, a_2, a_3, \ldots, a_m, \ldots, a_n)$ which satisfied the condition that for every positive ϵ there was an N_ϵ such that $| a_m - a_n | < \epsilon$ for all $m \geq N_\epsilon$, $n \geq N_\epsilon$. Such sequences satisfy Cauchy's condition for convergence, and hence are sometimes today called *Cauchy, regular* or *convergent*. Perron calls this "the criterion of Bolzano-Cauchy-Cantor."

Cantor considered that every sequence of this type represented a *real* number. It is fairly easy to show that this new set of numbers, defined and represented by limits of regular sequences, contains within it at least one sequence for each rational number. For example the sequence, $(1.9, 1.99, 1.999, \ldots, \ldots)$ formed in the obious way corresponds to 2, while $(.6, .66, .666, \ldots \ldots)$ corresponds to 2/3.

Further, the sum of two regular sequences $(a_1, a_2, a_3, \ldots, a_n, \ldots) + (b_1, b_2, b_3, \ldots, b_n, \ldots)$ is defined to be $(a_1 + b_1, a_2 + b_2, a_3 + b_3, \ldots, a_n + b_n \ldots)$. This sum can be shown not only to possess the usual desirable properties of associativity and commutativity, but also demonstrates that the sum of the sequences representing two rationals will turn out to be a sequence representing that rational which is the sum of the two original rationals.

After similar definitions of the product of two sequences, it can be shown that if sequences of rationals are thought of as elements of a set, then this set, with the operations noted above, forms a number *field*. That is, addition and multiplication are closed, associative, and commutative; there is

an identity element with respect to each operation; each element, except possibly zero, has an additive and multiplicative inverse; and multiplication is distributive with respect to addition.

Not only is the set of all regular sequences a field, but, as suggested above, there is a subset of these sequences which can be identified with the set of all rational numbers. Further, this subset also satisfies the field axioms in such a way that the result of adding or multiplying two rationals corresponds to the sequence which would result from adding or multiplying the sequences associated with the two original rationals. This isomorphism of the set of rationals with a subset of the set of regular sequences shows that the set of sequences is an *extension field* of the field of rationals.

Of course, just as there are many rationals corresponding to the integer 2 (e.g., 2/1, 4/2, 6/3, etc.), so there are many sequences corresponding to the rational 2/1. For example, in addition to that cited above, both $(1, 3/2, 7/4, 15/8, \ldots)$ and $(2, 2, 2, 2, \ldots)$ are such sequences. All of these are equal, however, under the definition of equality for regular sequences.

Before proceeding to a brief discussion of Dedekind's approach to irrationals, we should note several results related to Cantor's work. For example, the final solution of the ancient Greek problem of squaring the circle (with compasses and unmarked straight edge) was ultimately achieved only when Lindemann showed pi to be a transcendental number. Hence the problem is impossible of solution since ruler and compasses can construct only certain types of algebraic irrationals.

Euler had shown e to be irrational in 1737, long before Cantor; and Lambert had then shown pi to be irrational in 1767. However, some irrationals, such as the $\sqrt{2}$, are constructible, others such as $\sqrt[3]{2}$, cannot be so constructed, and in particular none of that class of irrationals called "transcendental"[3] can be so constructed.

Although their existence had long been suspected, it was not until 1844 that Liouville proved that there are such things as transcendental numbers. Even then he did not produce a single particular example. In 1871 Cantor proved that there is an infinite number of transcendentals. To do this, his theory of transfinite numbers and infinite aggregates was needed. This theory showed as well that, although the transcendentals and rationals are both infinite in number, there are actually more transcendentals than there are rationals or even algebraic numbers.

The first number to be proven transcendental was e. Hermite demonstrated this in 1873. Lindemann's proof of the transcendence of pi followed in 1882.

In spite of the large number of transcendentals proven to exist, it has been and still is a difficult problem actually to identify them. In fact, the seventh of the famous twenty-three unsolved problems presented by David Hilbert in 1900 was to prove the transcendence of a class of numbers, of which $2^{\sqrt{2}}$ and e^{π} were examples. In 1929, the Russian, A. Gelfond, proved that e^{π} is transcendental, and in 1934 he gave a complete solution for Hilbert's problem, but the testing of particular numbers for transcendence is still not easy.[4]

Dedekind's extension of the rationals also defined new numbers by use of sets of the "old" rational numbers. However, instead of discussing convergent sequences, he talked of partitions of the set of all rationals into two sets, such that every number of the second set was greater than every member of the first. Every such partition was called a *schnitt*, or a *cut*, and by definition was identified with a real number.

In some cuts there is a last element in the first set, or a first element in the second. Cuts of this type are identified with the rational numbers. But there are also cuts that have no last element in the first set, nor first in the second. Such cuts may also be regarded as defining real numbers, but in these cases the numbers correspond to irrationals.

For example, if the first set includes all the rationals whose square is less than 2, it has no last element, and the second set containing all rationals whose square is greater than 2 has no first element.

These two sets are a partition of the set of all rational numbers which defines the $\sqrt{2}$.

After defining addition, multiplication, and equality of cuts, they too can be shown to be a field which has a subset that is isomorphic to the rationals. Hence the set of cuts is an extension field of the rationals.

In teaching, there are two other related topics that ought to be mentioned with the study of incommensurables and irrationals. First, every rational number can be written as a periodic or repeating decimal fraction, while irrationals correspond to infinite nonrepeating decimals: Second, "quadratic surds" correspond to repeating continued fractions. The first fact should be the topic for an interesting discussion in many secondary school classes because the period of the decimal representation of rationals is associated with the use of the base 10, while the correspondence itself may be associated with geometric progressions in second-year algebra. Continued fractions are a less common topic which merit more attention as enrichment material associated with rationalizing radical expressions and the Euclidean algorism.

NOTES

1. P. S. Jones. "Irrationals or Incommensurables I, II, III, IV." *The Mathematics Teacher*, XLIX (February, March, April, October, 1956), pp. 123–127, 187–191, 282–285, 469–471).

2. Oskar Perron, *Irrationalzahlen* (1939), p. 55 ff. Although it has not been quoted directly, much use in this series has also been made of notes of J. Molk's translation exposition of A. Pringsheim's "Nombres Irrationnels et Notion de Limite" in Tome I, vol. 1 of *Encyclopédie des Sciences Mathématiques Pures et Appliquées* (Paris, 1904).

3. Transcendental numbers are those which cannot be the root of any algebraic equation. Algebraic equations are all those which can be written in the form $a_n x^n + a_{n-1} x^{n-1} + \ldots ax + a = 0$ where the exponents are integers and the a_i are rational numbers.

*4. Harry Pollard, *The Theory of Algebraic Numbers* (Mathematical Association of America, 1950) pp. 42–46; Einar Hille, "Gelfond's Solution of Hilbert's Seventh Problem," *American Mathematical Monthly*, XLIX (December, 1942), 654–661.

---

*Available as a Dover reprint.

15

The Genesis of Point Set Topology: From Newton to Hausdorff

JEROME H. MANHEIM

$\mathcal{P}$ARADOXES OF THE INFINITE, discussed by the Greeks, became urgent problems with the codification of calculus. Mathematicians of the seventeenth and eighteenth centuries sought to refer analysis to arithmetic in a way that would allow them to retain their reliance on the visual sense, i.e., on geometrical representation. During this period, when Euclidean geometry was the final truth-referent in mathematics, such an approach was clearly indicated. Berkeley's valid objections (1734) to the processes of Newton and Leibniz were not supplemented with a constructive program. The only possible constructive program, the arithmetization of analysis, would have been most unlikely as long as geometry occupied the dominant role. The works of the early analysts are viewed as the point of departure for systematizing problems which led to the development of point set topology.

The movement towards abstraction and generalization in mathematics is a measure of progress towards the notion of a topological space. In this respect a first debt is owed to Euler (1755) and other formalists, men who sought to work beyond the confines of geometry. But eighteenth-century formalism was a failure because it developed without adequate attention being paid to questions of convergence, and thereby yielded results which were demonstrably false.

Ultimately analysis depended upon a satisfactory theory of limits. Recognition of this fact is attributable to D'Alembert (ca. 1750), while credit for its implementation at what might be called the "naive level" belongs to Cauchy (1821). Mathematicians came to realize that an acceptable structure of analysis made the same demands upon rigor as did geometry and this gave rise to two separate, conflicting, lines of inquiry. The one typified by D'Alembert (1754), Gauss (1812), and Cauchy (1821) considered the limit concept as basic, while men such as Maclaurin (1742), Landen (1764), and Lagrange (1813) sought to achieve rigor by eliminating limit considerations from analysis. Cauchy, in particular, believed that the limit concept, by replacing the vague notion of infinitesimals, would be the instrument which would banish geometric intuition from analysis. Lagrange applied his skill in an effort to achieve a rigorous construction within the confines of the discrete. Although each school achieved a measure of success, neither fulfilled its ambition. Those who sought to employ the limit concept failed to make a valid definition. In essence their definition of a limit depended upon the prior existence of the same limit. Alternatively, the would-be arithmetization implicitly employed the limit notion.

Reprinted from *Mathematics Teacher* 59 (Jan., 1966): 36–41; with permission of the National Council of Teachers of Mathematics. Originally published as Section 7.1 of the final chapter of Jerome H. Manheim, *The Genesis of Point and Set Topology* (Oxford: Pergamon Press; and New York: the Macmillan Company, 1964). Permission to reproduce this section was kindly granted by Pergamon Press.

While the movement towards arithmetization had its beginnings in attempts to rigorize calculus, a major impetus was given to this tendency by the growing realization that the possibility of geometrical representation was inadequate as a base for mathematical truth. Problems that arose as a result of the analysis of the differential equation of the vibrating string caused new consideration to be given to Euler's contention of the inadequacy of geometry. Euler himself continued to play the role of formalist in his considerations of the string problem. This time, however, his manipulations were coupled with exceptional insight into the physical situation. Thus, while Euler (1750) identified a function with its graph, he felt that truth was more apt to be uncovered by formal processes than by reliance on the visual sense. D'Alembert's (1749) notion of function was indeed more general than Euler's, but when limited by the then-current belief that the graph of all functions was a proper subset of all graphs, it became a very restrictive concept. From the point of view of the development of topology, the importance of both these views of functions rests in their destruction.

The broadening of the function concept had its roots in Daniel Bernoulli's solution of the vibrating string problem (1750). Had Bernoulli's argument for a series solution been more mathematical and less speculative, an extension of the notion of function clearly would have been indicated. Instead, Euler gave a counter argument (1755) which was intended to demonstrate the absurdity of Bernoulli's claim to generality, an argument which preserved Euler's concept of function. Lagrange came close to validating Bernoulli's result and deriving the (Fourier) series expansion for very general functions (1759). He failed largely because of his single-minded purpose to prove the correctness of Euler's conclusions.

Fourier, in the course of his investigations on the theory of heat (1807), made the claim: "But herein we have dealt with a single case only of a more general problem, which consists in developing any function whatever in an infinite series of sines or cosines of multiple arcs."[1] Since to Fourier a function included arbitrary representations as well as those given by formulas, this statement initiated numerous inquiries intended to establish the exact reach of its validity. These investigations gave rise to the study of point set theory.

The first rigorous study of Fourier series was undertaken by Dirichlet in 1823. Here, for the first time, sufficiency conditions were soundly proved. This study confirmed a certain generality for Fourier representability and furthermore demonstrated, by counter-example, that the inclusiveness asserted by Fourier (1807) was false. It was in the course of these inquiries that Dirichlet (1837) extended the function concept, specifically the notion of continuous function, to include very general correspondences.

Riemann (1854) recognized the desirability of establishing necessity conditions that a function be Fourier representable. Since the coefficients of such an expansion are found by integration, Riemann first entered into an investigation of the definite integral, extending the meaning to certain functions which do not satisfy the piecewise continuity requirement laid down by Cauchy (1821). The fact that certain infinite sets of discontinuities did not destroy the integrability property, while others did, was a first step towards a classification of such sets. Riemann, however, limited his generalization of the integral concept to the study of Fourier series and did not pursue the set-theoretical consequences. Many of his results were employed by later researchers who sought to establish conditions for the uniqueness of the Fourier representation.

The relative status of geometric and analytic representation was not demonstrably disturbed by the introduction of infinite point sets into Fourier series considerations. The heirs of Fourier believed they had established the equal potency of the two forms of representation and were satisfied that this was the final determination. That there might be functions analytically expressible whose graphs would not be sketched was not even imagined.

Bernard Bolzano (1817) saw the necessity of proving the Fundamental Theorem of Algebra without reference to geometric intuition. His demonstration ultimately depended upon establishing the sufficiency of Cauchy's convergence criterion. Cauchy's supposed analytic proof of this theorem (1821) did not suppress, but simply altered, the role of geometric intuition. Bolzano's failure was logical in nature.

Bolzano gave a method for constructing a continuous curve which was not differentiable at an everywhere dense set of points (ca. 1830), but he did not publish his example. It was unlikely, even had it been published, that it would have had any significant effect on the mathematical community; since no formula was given. Bolzano's curve would not have been considered a function.

An example of a continuous function with an everywhere dense set of points of non-differentiability was given by Riemann in his 1854 *Habilitationsschrift*. But this example (which by all criteria then in effect was surely a function) failed to excite much interest in either Riemann or his distinguished audience, and remained unpublished until 1867.

The method of Riemann was generalized by Hankel (1870) and Cantor (1882). The importance of pathological functions, as they affect the foundations of analysis, resulted from a discovery by Weierstrass (1861). His remarkable function, which was everywhere continuous and nowhere differentiable, impressed upon mathematicians the necessity of restructuring all of analysis without reference to spatial intuition. This recognition was a clear vindication of the need expressed by Bolzano half a century earlier. The programs which began as an attempt to rescue analysis from the onslaught of pathological functions led to a new discipline, point set topology, a subject ideally suited to the study of just such functions.

To rebuild the foundations of analysis it was necessary to obtain a satisfactory definition of irrational numbers from the rationals. Four mathematicians were associated with this process.

Charles Meray (1869) saw that the definition of an irrational number as the limit of a certain sequence of rational numbers, as given by Cauchy in 1821 and in use since then, must rest upon geometric intuition if it was to avoid circularity. In Meray's reconstruction of the real numbers he begins with the notion of *progressive variable*. A progressive variable v is a quantity (there is no connotation of measurement in this use of the term) that receives an infinite set of values $\{v_n\}$, these values being taken from the domain of the rationals. Since the rational numbers are given, there is no difficulty in assigning a meaning to *rational limit*. It follows that if V is a rational limit of $\{v_n\}$ then, for n sufficiently large and for arbitrary $\in > 0$, $|v_{n+p} - v_n| < \in$ for all p. If the inequality still holds when there does not exist a rational limit, the sequence is said to have a *fictitious limit*, and these fictitious limits are called *fictitious numbers*. The fictitious numbers thus created are the irrationals.

Weierstrass (ca. 1870) defined a *numerical quantity* as a set of elements subject to the conditions that each element may occur only a finite number of times and the number of times each element occurs is known. The value assigned to a numerical quantity with a finite number of elements is the sum of its elements. After defining the meaning to be attached to the sum of finitely many numerical quantities, Weierstrass extends the definition to include infinitely many such objects. This allows consideration to be given to the sum of a convergent infinite set of (rational) numerical quantities. Those sums that are not rational define the irrationals.

Cantor (1872) defines an *elementary series* as a sequence of rational numbers $\{a_i\}$ such that a_n and $a_n - a_{n+r} \to 0$ with $1/n$. Such a sequence may or may not have a rational limit. But whether or not it has such a limit the series serves to define a real number. If there is no rational limit, the real number so defined is called irrational.

Dedekind (ca. 1858) partitioned the rationals into two nonempty subsets A and B such that every rational is in A or B and such that every member of A is less than every member of B. Respecting these sets and the phrases "has a last element," "has a first element," there are four permutations:

1. *A* has a last element and *B* has a first element.
2. *A* has no last element but *B* has a first element.
3. *A* has a last element but *B* has no first element.
4. *A* has no last element and *B* has no first element.

Situation (1) is impossible by virtue of the density property of the rationals. Numbers (2) and (3) serve to define the rational which is the first element of *B* or the last element of *A* respectively. The fourth situation, which does not identify a rational, is used by Dedekind to define the irrationals.

There were numerous objections, both to arithmetization *per se* and to the methods employed by the arithmetizers. But the majority of mathematicians became convinced that analysis was firmly embedded in the integers. With reliance upon spatial intuition eliminated as a necessary component of analytic investigations, it became possible to study hitherto proscribed topics. When abstract spaces made their appearance some years later, no significant objections could be raised. This generalized notion of spaces, however, depended also on a theory of point sets.

Point set considerations developed from attempts to establish conditions under which the representation of a function by a trigonometric series is unique. The need to establish uniqueness arose from recognition of the role of uniform convergence in Fourier series investigations. Abel (1826) had given an example of a convergent sequence of continuous functions which failed to converge to a continuous function. This example had little impact on the development of mathematics until Seidel related this phenomenon to uniform convergence in 1848. Weierstrass demonstrated that if a function had a series expansion which converged uniformly, then the sum of the integrated terms represented the integral of the function. In 1870 Heine noted that the proof that a function had a unique Fourier expansion depended upon termwise integration and hence tac-

itly assumed uniform convergence. This observation inaugurated a sequence of investigations designed to establish the conditions under which the Fourier expansion was unique, a question of practical as well as theoretical importance. The initial results of point set theory were a by-product of these investigations.

The first attempt to rescue Fourier series from what appeared as the destructive influence of uniform convergence was due to Heine (1870). He showed that if a function satisfied the Dirichlet conditions in $(-\pi, \pi)$ its Fourier series is uniformly convergent in any open interval if the function is continuous throughout the interval and at the end points.[2] Four years later Paul du Bois-Reymond showed that continuity of a function was insufficient to guarantee the uniform convergence of its series so that, even for a continuous function, there may be more than one trigonometric series.

The uniqueness problem engaged the interest of Georg Cantor. He began his researches by considering the most general trigonometric series, i.e., a series whose coefficients are not restricted to the integral form. In his papers, which started to appear when he was only twenty-five years old (1870), he established several important theorems. One of these results (1872) showed that an infinite set of points of discontinuity was not, of itself, sufficient to destroy uniqueness, a result complementary to that of du Bois-Reymond (1876). In the course of proving this theorem Cantor found it necessary to distinguish between various types of infinite point sets. Distinctions made by earlier investigators (with the exception of du Bois-Reymond) had invariably been made on the basis of isolating a subset (not necessarily proper) of the rationals from the reals. This type of classification failed to suit Cantor's needs and, what is more surprising, under his arrangement into *kinds* both the rationals and the reals were of the same kind.

Cantor saw that a theory of sets had implications for all branches of mathematics and, as had been anticipated by Dirichlet, that the set concept was more fundamental than that of function. The subsequent development of naive set theory is due almost exclusively to Cantor (1872 *et seq.*). Most mathematicians insisted, contra Cantor, that infinity was only approachable and consequently not subject to an arithmetic. Eventually the importance of Cantor's approach was recognized and, while one group of mathematicians was concerned with eliminating the antinomies of set theory through axiomatization, another reached for generalization.

Throughout the evolution of the function concept geometry had been under attack, but some mathematicians refused to abandon this subject simply because it lost its supremacy to analysis. These men sought to apply concepts which were essentially nongeometric to the classification of geometries. A first attempt was made by Riemann in 1854. Eighteen years later Felix Klein, in his famous *Erlanger Programm*, considered that he had brought order to the classification of geometries via the theory of groups. But the *Erlanger Programm* failed to recognize that a space, by itself, can be an object of study, independent of the nature of the transformations of its elements. The emergence of point set topology demonstrated the inadequacy of Klein's view.

The study of sets of objects other than points began in 1833 with Ascoli's paper on sets of curves. Shortly thereafter (1887) Volterra and Arzela studied sets of functions, researches which gave rise to Functional Analysis. Also in 1887 Hadamard proposed that investigations be made of restricted function sets. Borel (1903) suggested a way of extending the notion of *nearness*, and therefore *limit*, to sets of lines and planes. In this way he generalized many of the concepts originally developed for point sets by Cantor. Fredholm's work of the same year led to the notion of a set of functions as a set of points. About this time Hilbert (1901) considered the set of functions for which the Bolzano-Weierstrass theorem held true.

The generalization by abstraction of the set concept was made by Fréchet. In his 1906 thesis he considered first an arbitrary set of elements E and an operation which maps E into some numerically determined value. The generality Fréchet afforded to the set of relations involving the points

meant that the relations were not necessarily explainable in terms of the group concept. Fréchet's work, based upon limit, did not, however, achieve the generality of a topological space.

Riesz based his treatment of sets on the concept of condensation point (1908), hoping thereby to avoid the limitations of Fréchet's study, but he did not develop many of the consequences. In 1913 Hermann Weyl proposed a development of general two-dimensional manifolds from the point of view of neighborhoods. For all practical purposes Weyl was an axiom away from the notion of a topological space, an axiom of separation.

The appearance in 1914 of Hausdorff's *Grundzüge der Mengenlehre* marks the emergence of point set topology as a separate discipline. A *space* became merely a set of points and a set of relations involving these points, and a geometry was simply the theorems concerning the space. Hausdorff began his development of topology with a small noncategorical set of neighborhood axioms. After deriving many of the properties of the general space thus defined he progressively introduced new axioms, ultimately deducing metric spaces and, finally, particular Euclidean spaces.

Looking backwards, point set topology developed from arithmetization and set theory. The latter owes its origin to problems in Fourier series representation while the former is principally indebted to the introduction of pathological functions into analysis. Both these problems, in turn, had their roots in the notion of function, a notion which for long sought to preserve the dominant role of geometry. The evolution of the function concept began with the codification of calculus, when the apparant paradoxes of the infinitesimal processes became serious problems.

NOTES

1. Joseph Fourier, *The Analytical Theory of Heat*, trans. Alexander Freeman (New York: Dover, 1955), p. 168.

2. Horatio Scott Carslaw, *Introduction to the Theory of Fourier's Series and Integrals* (New York: Dover, 1930), p. 13., notes that it was shown later that a Fourier series could be integrated termwise even if the series did not converge.

The Origin of the
Four-Color Conjecture

KENNETH O. MAY

*C*ONSIDERING THE FAME and tender age of the four-color conjecture,[1] our knowledge of its origins is surprisingly vague. A well-known tradition appears to stem from W. W. R. Ball's *Mathematical Recreations and Essays,* whose first edition appeared in 1892. There it is said that "The problem was mentioned by A. F. Möbius in his lectures in 1840 . . . but it was not until Francis Guthrie communicated it to De Morgan about 1850 that attention was generally called to it. . . . It is said that the fact had been familiar to practical mapmakers for a long time previously."[2] In spite of repetition by later writers, this tradition does not correspond to the facts.

In the first place, there is no evidence that mapmakers were or are aware of the sufficiency of four colors. A sampling of atlases in the large collection of the Library of Congress indicates no tendency to minimize the number of colors used. Maps utilizing only four colors are rare, but those that do so usually require only three. Books on cartography and the history of mapmaking do not mention the four-color property, though they often discuss various other problems relating to the coloring of maps.

If cartographers are aware of the four-color conjecture, they have certainly kept the secret well. But their lack of interest is quite understandable. Before the invention of printing it was as easy to use many as to use few colors. With the development of printing, the possibility of printing one color over another and of using such devices as hatching and shading provided the mapmaker with an unlimited variety of colors. Moreover, the coloring of a geographical map is quite different from the formal problem posed by mathematicians because of such desiderata as coloring colonies the same as the mother country and the reservation of certain colors for terrain features, e.g., blue for water. The four-color conjecture cannot claim either origin or application in cartography.

To support his statement about Möbius, Ball refers to an article by Baltzer, a former student of Möbius and the editor of his collected works.[3] However, as has been pointed out by H. S. M. Coxeter,[4] this article shows merely that Weiske communicated to Möbius a puzzle whose solution amounted to the claim that it is impossible to have five regions each having a common boundary with every other. Baltzer witnesses that in 1840 Möbius presented this puzzle to a class and laughingly revealed its impossibility in the next lecture. But nothing was published on the problem, and its

Reprinted from *Mathematics Teacher* 60 (May, 1967): 516–19; with permission of the National Council of Teachers of Mathematics. This article is based on work done during the tenure of a Science Faculty Fellowship from the National Science Foundation. It was presented to the Minnesota Section of the Mathematical Association of America on November 3, 1962, and to the Midwest Junto of the History of Science Society on April 5, 1963. It appeared in *Isis,* LVI (1965, 3, No. 185), pp. 346–48, and is republished here with the kind permission of that journal. The author is indebted to S. Schuster, G. A. Dirac, J. Dyer-Bennet, Oystein Ore, and H. S. M. Coxeter for discussion and suggestions.

source remained unknown until Baltzer discovered Weiske's communication among the Möbius papers more than forty years later, when the four-color conjecture was already well known. Baltzer found no evidence that Möbius worked on the problem, and it is not mentioned in the collected works published in 1885–1887. Evidently Weiske's puzzle and its mention by Möbius were fruitless and had no historical link with the origin of the four-color conjecture.

The statement that attention was generally called to the problem about 1850 appears also to be not entirely accurate. Indeed, the first printed reference to it appeared in 1878, when the *Proceedings of the London Mathematical Society* reported Cayley's question as to whether the conjecture had been proved.[5] Interest was immediate, and a long series of partial solutions and pseudosolutions began with the papers of Kempe[6] and Tait.[7] Cayley, Kempe, and Tait attributed the problem vaguely to Augustus De Morgan.

More precise information was supplied by the physicist Frederick Guthrie in a communication to the Royal Society of Edinburgh in 1880. He wrote,

> Some thirty years ago, when I was attending Professor De Morgan's class, my brother, Francis Guthrie, who had recently ceased to attend them (and who is now professor of mathematics at the South African University, Cape Town), showed me the fact that the greatest necessary number of colors to be used in coloring a map so as to avoid identity of color in lineally contiguous districts is four. I should not be justified, after this lapse of time, in trying to give his proof, but the critical diagram was as in the margin. [The diagram shows four regions in mutual contact.]
>
> With my brother's permission I submitted the theorem to Professor De Morgan, who expressed himself very pleased with it ; accepted it as new; and as I am informed by those who subsequently attended his classes, was in the habit of acknowledging whence he had got his information.
>
> If I remember rightly, the proof which my brother gave did not seem altogether satisfactory to himself; but I must refer to him those interested in the subject. I have at various intervals urged my brother

to complete the theorem in three dimensions, but with little success. . . .[8]

A hitherto overlooked letter from De Morgan to Sir William Rowan Hamilton permits us to pinpoint events more precisely than could Frederick Guthrie, after the passage of almost thirty years. On October 23, 1852, De Morgan wrote as follows:

> A student of mine asked me today to give him a reason for a fact which I did not know was a fact, and do not yet. He says, that if a figure be anyhow divided, and the compartments differently coloured, so that figures with any portion of common boundary *line* are differently coloured—four colours may be wanted, but no more. Query cannot a necessity for five or more be invented? As far as I see at this moment, if four *ultimate* compartments have each boundary line in common with one of the others, three of them inclose the fourth, and prevent any fifth from connexion with it. If this be true, four colours will colour any possible map, without any necessity for colour meeting colour except at a point.
>
> Now, it does seem that drawing three compartments with common boundary, two and two, you cannot make a fourth take boundary from all, except inclosing one. But it is tricky work, and I am not sure of all convolutions. What do you say? And has it, if true, been noticed? My pupil says he guessed it in colouring a map of England. The more I think of it, the more evident it seems. If you retort with some very simple case which makes me out a stupid animal, I think I must do as the Sphynx did. If this rule be true, the following proposition of logic follows:
>
> If *A, B, C, D,* be four names, of which any two might be confounded by breaking down some wall of definition, then some one of the names must be a species of some name which includes nothing external to the other three.[9]

On October 26, Hamilton replied: ". . . I am not likely to attempt your 'quaternion of color' very soon." Evidently the response of others to the conjecture was equally passive. Even Francis Guthrie published nothing on it, though he lived until 1899 and produced a book and several papers on other topics.

Apparently it was on October 23, 1852, that Frederick Guthrie communicated his brother's conjecture to De Morgan, who did not bother to explain to Hamilton the indirect nature of the communication.[10] Probably De Morgan, in telling others about the problem, mentioned that it had occurred to Guthrie while coloring a map, and this gave rise to the tradition linking the conjecture with the experience of cartographers.[11]

On the basis of the data given here we can replace tradition by the following account of the origin of the four-color conjecture. It was not the culmination of a series of individual efforts, but flashed across the mind of Francis Guthrie, a recent mathematics graduate, while he was coloring a map of England. He attempted a proof, but considered it unsatisfactory—thus showing more critical judgment than many later workers. His brother communicated the conjecture, but not the attempted proof, to De Morgan in October 1852. The latter recognized the essentially combinatorial nature of the problem, gave it some thought, and tried without success to interest other mathematicians in attempting a solution. He communicated it to his students, giving due credit to Guthrie, and to various mathematicians, one of whom revived the problem almost thirty years later and launched it on its erratic course.

Rarely is a mathematical invention the work of a single individual, and assigning names to results is generally unjust. In this case, however, it would seem that the four-color conjecture belongs uniquely to Francis Guthrie and could fairly be called Guthrie's problem.[12]

NOTES

1. In nontechnical terms the four-color conjecture is usually stated as follows: Any map on a plane or the surface of a sphere can be colored with only four colors so that no two adjacent countries have the same color. Each country must consist of a single connected region, and adjacent countries are those having a boundary line (not merely a single point) in common. The conjecture has acted as a catalyst in the branch of mathematics known as combinatorial topology and is closely related

to the currently fashionable field of graph theory. More than half a century of work by many (some say all) mathematicians has yielded proofs only for special cases (up to 35 countries in 1940). The consensus is that the conjecture is correct but unlikely to be proved in general. It seems destined to retain for some time the distinction of being both the simplest and most fascinating unsolved problem in mathematics.

* 2. W. W. Rouse Ball, *Mathematical Recreations and Essays*, rev. H. S. M. Coxeter (London: Macmillan & Co., 1959), p. 223.

3. R. Baltzer, "Eine Erinnerung an Möbius und seinen Freund Weiske," *Bericht über die Verhandlungen der Sächsischen Akademie der Wissenschaften zu Leipzig. Math.-Nat. Kl.* XXXVII (1885), 1–6.

4. H. S. M. Coxeter, "The Four-Color Map Problem," *The Mathematics Teacher*, LII (April 1952), 283–89.

5. A. Cayley, "On the Colouring of Maps," *Proceedings of the London Mathematical Society*, IX (1878), 148. A four-line report of Cayley's query in the session of June 13. See also a note with the same title by Cayley in the *Proceedings of the Royal Geographical Society*, I (1879, N.S.), 259–61.

6. A. B. Kempe, "On the Geographical Problem of Four Colors," *American Journal of Mathematics* II (1879), 193–204.

7. P. G. Tait, "On the Colouring of Maps," *Proceedings of the Royal Society of Edinburgh*, X (1880), 501–3, 729.

8. Frederick Guthrie, "Note on the Colouring of Maps," *Proc. Roy. Soc. Edinburgh*, X (1880), 727–28.

9. R. P. Graves, *Life of Sir William Rowan Hamilton* (Dublin, 1889), III, 423.

10. Biographical data on the Guthrie brothers supports the date indicated by De Morgan's letter. Indeed, Francis took his B.A. at University College London in 1850 and his LL.D. in 1852, whereas his younger brother Frederick was a student there in 1852 and, after a period in Germany, received his bachelor's degree in 1855.

11. In his note of April 1879, referred to in Note 5 above, Cayley begins: "The theorem that four colors are sufficient for any map, is mentioned somewhere by the late Professor De Morgan, who refers to it as a theorem known to map-makers." I have been unable to locate this "somewhere" and suspect that the communication was verbal. The same vague reference to De Morgan appears in a note in *Nature*, XX (1879), 275, and in Kempe's article cited in Note 6.

12. It is so called in E. Lucas, *Récréations mathématiques* (Paris, 1894).

*Available as a Dover reprint.

Topological Diversions

Topology is the study of the properties of geometric shapes which do not change when the shapes themselves are deformed through stretching or twisting. Size and distance (length) are irrelevant in topology. Topological transformations such as twisting can result in some interesting objects that possess unusual features. A. F. Möbius (1790–1868) discovered a geometric configuration containing only one side and one surface edge. It is called a Möbius strip in honor of him. A Möbius strip can easily be constructed by taking a thin strip of paper, twisting it once, and joining the two ends together:

In 1882, the German mathematician Felix Klein extended the concept of a Möbius strip into a "bottle" which had no inside nor outside or edges. An anonymous verse celebrates this event:

> A mathematician named Klein
> Thought the Möbius band was divine.
> Said he: "If you glue
> The edges of two,
> You'll get a weird bottle like mine."

For dramatic displays of a Klein bottle and its properties conduct an Internet search on the topic.

17

Meta-Mathematics and the Modern Conception of Mathematics[1]

MORTON R. KENNER

*I*T IS PERHAPS PERTINENT to begin this article with a few words indicating why it is proper that we who are concerned with mathematics education should also be concerned with such things as meta-mathematics. In this journal, it is hardly necessary to belabor the well-known facts that secondary school and college enrollments have swelled and are continuing to swell, that secondary and undergraduate curricula are being altered, revised, and experimented with. It should also be obvious, from the pages of this journal, that as the facilities of our colleges become more and more overtaxed and as our high school populations overflow into junior colleges—in many cases identical to the high school—our secondary teachers, and hence our secondary teacher training program, of necessity, will have to become cognizant of changes which have occurred in mathematics somewhat beyond and apart from the classically conceived secondary level. These facts create many problems about which we are all aware. I do not intend to try to solve any of them in this article. But, whatever the solutions to these new problems will be—and there will undoubtedly be many solutions—it seems to me that all of these solutions will share at least one thing in common. All of these solutions will in some sense recognize the contemporary view of the abstract nature of mathematics. They will all recognize that mathematics is a good deal more

Reprinted from *Mathematics Teacher* 51 (May, 1958): 350–57; with permission of the National Council of Teachers of Mathematics.

than how much or how many, a good deal more than do this and don't do that.

Sharing this contemporary view of the nature of mathematics does not necessarily mean that secondary mathematics must be conceived of and taught entirely abstractly—although it may well mean that. But it does mean that we must ourselves, as teachers, know many of the consequences of believing that mathematics is abstract. It should be unnecessary to remind ourselves that it is a poor teacher indeed who teaches everything he knows. In fact, what so often makes excellent teaching excellent is that which we know but which we do not teach. Understanding what meta-mathematics is, to a certain extent, is equivalent to understanding what we mean when we say that mathematics is abstract. In the twentieth century, this latter understanding is certainly an obligation of all mathematics teachers.

The following discussion will attempt to sketch first some of the background necessary to an understanding of what meta-mathematics is, and second to discuss intuitively the nature of the subject and how it is related to abstractive tendencies in mathematics.

Mathematics and Abstraction

But what do we really mean when we say that mathematics is abstract? It is quite probable that we may mean two essentially different things. On the one hand, we may mean by abstractness the sort of thing which enables us to discuss a property (or properties) of a host of particular mathematical systems at the same time. We can—and do—talk about an abstract group, for example, knowing that in some sense we are also talking about certain aspects of the set of symmetries of a triangle, or about certain aspects of the set of rational numbers under addition, or about certain aspects of many other mathematical systems that are quite different in many ways but yet share the one property of being models or representations of an abstract group. This type of abstractness—the type which enables us to deal so efficiently with so many mathematical systems simultaneously by considering universally conceived properties shared by all of them—is an important part of current mathematics. To many, it represents in a nutshell what we mean when we say mathematics is abstract.

There is, however, another equally important meaning of abstractness. This other meaning of abstractness is most easily associated with the intuitive awareness which we display in recognizing that mathematics is independent—in some sense—of empirical knowledge. When we feel that the theorems of mathematics are not verified by experience or are not made true by experience or are not about experience, we are sensing this other meaning of abstractness. This is the side of abstractness which led to Mr. Russell's now famous quip about our ignorance. The evolution of this type of abstractness is common knowledge in the field of geometry where the development of non-Euclidean geometry, from at least one point of view, purports to show that geometry is not a physical science. This, indeed, is one of the major lessons that we learn in studying the development of non-Euclidean geometry. But we should not forget that this evolution took place also in the field of algebra.[2] As late as 1770, Euler still believed that mathematics was the science of magnitudes. Number, for Euler, was conceived as being in the first and last instance an answer to the question "how many?" and in cases of extensive measure, "how much?"

At about this same date, William Frend, in his book, *The Principles of Algebra*, printed in 1796, refused [sic!] to accept either negative or imaginary numbers. Frend defended his position with the following words—words undoubtedly strange to our twentieth-century ears. He said,

> The ideas of number are the clearest and most distinct in the human mind; the acts of the mind upon them are equally simple and clear. There cannot be confusion in them. . . . But numbers are divided into two sorts, positive and negative; and an attempt is made to explain the nature of negative numbers by allusion to book-debts and other arts. Now, when a person cannot explain the principles of a science without a reference to metaphor, the probability is that he has never thought accurately upon the subject. A number may be greater or less than another; it may be added to, taken from, multiplied into, and divided by another number; but in other respects it is intractable; though the whole world should be destroyed, one will be one, a three will be a three; and no art whatever can change their nature. You may put a mark before one, which it will obey: it submits to being taken away from another number greater than itself, but to attempt to take it away from a number less than itself is ridiculous. Yet this is attempted by algebraists, who talk of a number less than nothing, of multiplying a negative number into a negative number and thus producing a positive number, of a number being imaginary. Hence they talk of two roots to every equation of the second degree . . . they talk of solving an equation which requires two impossible roots to make it soluble: they can find out some impossible numbers, which, being multiplied together produce unity. This is all jargon, at which common sense recoils.[3]

It was not until the middle of the nineteenth century that Peacock, Gregory, De Morgan, and Boole, in a series of books and papers, laid the basis for the modern abstract conception of algebra. The first of these men, Peacock, in his *Treatise on Algebra* of 1830, distinguished between arithmetic and algebra. He believed that in arithmetic, the symbols A and B represent or stand for integers or fractions. In algebra, on the other hand,

Peacock felt that these signs need not represent integers and, in general, do not do so. Similarly, Peacock felt that in arithmetic the + and − signs represent or stand for the familiar addition or subtraction, whereas in algebra, only the formal properties of possible operations are relevant to this symbolic algebra. Peacock argued that when formally defined operations are symbolically denoted, certain expressions or forms are obtained which are equivalent to other forms in virtue of the rules of the systems of symbols. The discovery of these equivalent forms, he felt, constituted the principal business of algebra.

Gregory further clarified this point in 1838, when he pointed out that symbolic algebra is the science concerned not with the combinations of operations defined by their nature (i.e., what they are or do), but concerned solely with the laws of combination to which these operations are subject.[4]

De Morgan, in 1839, took a further step and divided the algebra of Peacock and Gregory into two different types of algebra—that which De Morgan called technical algebra and that which he called logical algebra.[5] Technical algebra for De Morgan is the art of using symbols under regulations which, when this part of the subject is considered independently of the other, are proscribed as the definitions of the symbols. Technical algebra then is a formal procedure, an operation on signs or symbols obeying prescribed rules. Logical algebra, on the other hand, is the science which investigates the method of giving meaning to the primary symbols and of interpreting symbolical results. Logical algebra then is similar to the problem of finding concrete examples or models for the system spelled out by the technical algebra. Thus, in a certain sense, we might say that De Morgan's technical algebra is an example of the second type of abstractness discussed above, while logical algebra is an example of the first type.

The important point to be noted here is that the validity of deduction in a technical algebra cannot depend upon the material interpretation of the symbols of the calculus. Technical algebra was algebra in which symbols were used independently of the meaning which logical algebra might give to them later. To adopt this viewpoint of technical algebra means—at least to those concerned with the axiomatizing of branches of mathematics—that it must be possible to regard the symbols of a system completely divorced from any meaning subsequently or earlier put upon them.

This point is a crucial one, for suppose that the meaning of the word "line" or the meaning of the word "point" were really necessary to the axiomatization of Euclidean geometry. This would then mean that our axioms did not fully prescribe what is to be meant by line or by point. This would mean further that part of the meaning of point and line must lie outside of the axioms.

Hilbert recognized this crucial point. His *Foundations of Geometry* was in large part an attempt to formalize Euclidean geometry by giving no meaning to the thought things "point" and "line" which were not *completely* stipulated by the axioms. Hilbert, of course, did much more, but his axiomatizing of Euclidean geometry is a landmark in the development of mathematics primarily because it develops a complete (and very rich) mathematical system without utilizing any meaning apart from that given by the axioms. His work was undoubtedly motivated by historical precedent, he undoubtedly was thinking of a specific model; within the system, however, the "thought things" were given meaning only by the axioms.

The Price of Abstractness

But if we are to look only to the axioms, only to the system, how is it possible to know that our system is not contradictory? That is, how do we know that our axioms and the logical rules we assume to deduce theorems from them are not such that at some time it will be possible to deduce a theorem and its contradictory? To deduce, say, both $1 = 1$ and $1 \neq 1$?

Before considering the question of how we can know that our system is noncontradictory, let us pause to consider the consequences of being able to deduce both a theorem and its contradictory. According to the rules of logic which we use in mathematics, it follows that a proposition and its negation imply any proposition. In symbols: (P and ~ P)→Q. Thus, if we could deduce P and also could deduce ~ P from our axioms using the accepted logical canons, it would follow that any statement expressible in the symbolism would also be a true theorem.

Now it might seem that the way out of this difficulty is trivial. All we need do is exhibit a statement which is expressible in the symbolism of the system. Show that this statement is false, that is, show that it is not a true statement and cannot be true and hence not a theorem, and we would be home free. For, clearly, if we could deduce a statement and its contradictory, every statement would be true. We have exhibited one which is not true. Hence, clearly it must be impossible to deduce a statement and its contradictory, i.e., our system is not contradictory. We shall see presently that this would indeed be a way out—but it is far from trivial. Let us return to the question raised above concerning our ability to know whether or not a system is contradictory.

Before the nineteenth century, sense intuition had served as a guide in telling us that our axioms were not contradictory. Let us see precisely what this means. When, say, Euclid formulated certain axioms for plane geometry, he felt that these axioms were statements—true and eternal statements—about the geometry of the real world. Thus, the physical world was really a model for the axioms. Now, if our axioms were contradictory, that is, if they could give rise to a pair of contradictory propositions, then our model would have to give rise to a pair of contradictory intuitions. But since Euclid held—with obvious justification—that our sense experience of space was not contradictory, and further, since he held that this space was an exact model of the axioms, it followed that our axioms could not be contradictory. When, however, sense experience was no longer considered to be a valid guide for our technical algebra or technical geometry—in De Morgan's sense—we could not use sense experience as a model to test the validity of our axioms. New methods were necessary. If our system was to be entirely self-contained (independent of meanings attached to symbols), it was necessary to find a way to insure that no pair of contradictory propositions could arise—i.e., could be deduced.

In his *Foundations of Geometry*, Hilbert to some extent bypassed this problem. In that work, he used the well-known techniques of interpreting geometric statements algebraically—that is, Hilbert used algebra as a model for the geometry and showed how all of the geometric statements could be consistently interpreted in terms of algebraic elements and relations.[6] This, in a sense, is of course what we do in analytical geometry. But clearly, showing that all geometric statements are capable of algebraic interpretation does not establish the fact that the geometric axioms are noncontradictory. All it can possibly do is demonstrate that *if* algebra is noncontradictory, then geometry is also noncontradictory.

It is important to recognize that interpreting the axioms by means of a model (however well known) does not necessarily strike at the heart of the deduction problem. We referred earlier to the fact that (P and ~ P)→Q. We said then that if it were posssible to demonstrate the existence of a statement which could only be false, our axioms would be noncontradictory. For otherwise, every statement would be true. But clearly such a task is far from trivial. It is not enough to show that a given statement is false, it must also be shown that its denial *cannot* be deduced from our axioms. It is thus necessary to analyze carefully the logical procedures by which theorems are arrived at. We must, in other words, analyze our methods of proof including both original axioms and our logic of deduction. Only in this way will it be possible to prove that from a set of axioms (shown perhaps to be noncontradictory through models) it is not possible to deduce both a proposition and its negation as true theorems.

Hilbert was forced to consider this question in his paper of 1905, "On the Foundations of Logic and Arithmetic."[7] As we remarked above, Hilbert had used algebra (and hence elementary arithmetic) in establishing the consistency of geometry. That is, Euclidean geometry was noncontradictory if algebra was. But when he took up the question of arithmetic itself, there was no model to turn to. Hilbert stated this in the following way:

> Arithmetic is indeed designated as a part of logic and it is customary to pre-suppose in founding arithmetic the traditional principles of logic. But on the attentive consideration, we become aware that in the usual exposition of the laws of logic certain fundamental conceptions of arithmetic are already employed. For example, the concept of the aggregate, and in part also the concept of number. We fall thus into a vicious circle and therefore to avoid paradoxes a partly simultaneous development of the laws of logic and arithmetic is requisite.[8]

Thus, since certain fundamental aspects of arithmetic are already employed in logic, the simultaneous development of both is necessary.

It is clear from the above quotation that the paradoxes involving the theory of aggregates or set theory had just caused great concern in the mathematical world. It would take us far afield to analyze the various types of paradoxes resulting from an uncritical use of the concept of set—in particular, infinite sets.[9] Suffice it to remark that Hilbert felt that the set concept, so fundamental in logic and hence the foundations of arithmetic, must be subjected itself to a new type of critical analysis.[10]

Let us pause for a moment to sum up what we have already discussed. The development of abstraction in mathematics has led to the viewpoint that mathematical assertions, or mathematical systems, as mathematics have nothing in common with sense experience. Technical algebra, from the point of view of De Morgan and subsequently, is algebra in which meanings to be imparted to symbols must be specified by the axiom system without any reference outside the system. Mathematical symbols must derive their entire meaning

from the postulates of a system itself. If they do not, we have not completely axiomatized it. But having abandoned sense experience as a criterion in settling mathematical questions, it is no longer possible to establish the consistency—or non-contradictoriness—of our postulate systems by asserting that sense models of them give rise to no contradicting sense intuitions.[11] The sense models tell us nothing whatsoever about our formal system. But mathematics, if it is to lay claim to perfect rigor, must not give rise to contradictions. When we ask our students to accept certain sets of postulates for geometry or for arithmetic, we do so with the faith that these postulates do *not* give rise to contra-dictions. It is important to show that this faith is well founded.

Meta-Mathematics

To subject our axiom systems and our logic simultaneously to a critical analysis, Hilbert created a new discipline, the discipline of meta-mathematics. In a moment we shall consider more carefully what this new discipline was. Let us now see what the program was which Hilbert delineated for it.[12]

1. Meta-mathematics must enumerate all symbols used in mathematics and logic. These Hilbert called the fundamental symbols.

2. Meta-mathematics must denote uniquely all combinations of these symbols which occur as meaningful propositions in classical mathematics. These Hilbert called formulas.

3. Meta-mathematics must yield processes of construction which enable us to set up or to arrive at all formulas which correspond to the demonstrable assertions of classical mathematics.

4. Meta-mathematics must demonstrate in a finite combinatorial way that these formulas which correspond to calculable arithmetic can be demonstrated in accordance with 3 (above) if and only if the actual calculations of the mathematical assertions corresponding to the formulas results in the validity of the assertions.

It should be noted that item 4 is in essence a demand for a finite proof of consistency—i.e., of noncontradictoriness—of classical mathematics. But what sort of discipline can meta-mathematics be? What sort of discipline will be capable of enumerating all symbols, of enumerating all formulas of classical mathematics, of yielding a process of construction which will arrive at such formulas, and finally of giving an absolute proof of consistency?

The first thing about such a discipline which I hope is becoming obvious is that the *object* which meta-mathematics studies is mathematics itself. Let us see what this means by examining statements of mathematics and statements of meta-mathematics. A statement of mathematics is:

For every x, if x is a prime and if $x > 2$, then x is odd.

Statements in meta-mathematics would be:

"x" is a numerical variable.

"2" is a numerical constant.

"Prime" is a predicate expression.

"$>$" is a binary predicate.

Note that when meta-mathematics says "'x' is a numerical variable," it is referring to the symbol—sign—"x" as it is used in the mathematical statement. When it says that "'2' is a numerical constant," it is referring to the symbol—sign—"2" as it is used in the mathematical statement.

Or as another example, consider the mathematical statement

$2 + 3 = 5$.

A statement in meta-mathematics would be:

"$2 + 3 = 5$" is a formula.

And when meta-mathematics says that "'$2 + 3 = 5$' is a formula," it is referring to how the symbols or signs have been put together and are used in mathematics. Thus, meta-mathematics is a subject whose object of study is mathematics.

The meta-mathematician operates with three separate and distinct subject matters. First, there is the informal mathematical system such as the natural numbers, or the points and lines and their relations in plane Euclidean geometry. This subject matter is informal and intuitive. Second, there is the formal axiomatic treatment of the informal mathematical system. This, you will recall, is motivated by the informal mathematical systems but if the axiomatic treatment of this informal mathematical system is to be successful, the formal system must itself be completely independent of the informal system. This is, of course, an immediate consequence of saying that mathematics is abstract, or that mathematical truths are not dependent upon sense experience. And finally, there is the meta-mathematics. This is a discipline which itself must be carefully developed. It is to a certain extent also informal, but the informalities present in meta-mathematics are of a different type. They are informalities in the sense of establishing criteria for permissible procedures. For example, no infinite operations are permissible.

From the point of view of the meta-mathematics, the formal system is independent of any meaning which may be given to it by an informal mathematical system. Meta-mathematics sees only the symbols, how they are grouped together, what is the logical pattern of passing from one grouping of symbols to another grouping of symbols. At the meta-mathematical level, we may speak of this as deduction, but in the formal system, we must regard it solely as a symbolic process. From the point of view of the informal mathematical system, we may regard this process as the means of arriving at truths about our interpreted objects; but from the meta-mathematics, it appears solely as a procedure which either follows rules and hence can be called "deduction" or does not follow these rules and hence cannot be so called.

It seems proper, at this point, to present a brief example of meta-mathematics which may give more meaning to the preceding discussion.[13] Assuming that we had listed the symbols to be permitted and also the operation of putting them side by side, i.e., operation of juxtaposition, we might then proceed as follows to define "term."

1. 0 is a *term*.
2. A variable is a *term*.
3. If s and t are *terms*, the $(s) + (t)$ is a *term*.
4. If s and t are *terms*, then $(s) \cdot (t)$ is a *term*.
5. If s is a *term*, then $(s)'$ is a *term*.
6. The only terms are those given by 1–5.

THEOREM: $((c)') = (a)$ is a *term*.
PROOF: By 2 a and c are terms.
      By 5 $(c)'$ is a term.
      By 3 $((c)') + (a)$ is a term.

We could now define "formula."

1. If s and t are terms, the $(s) = (t)$ is a *formula*.
2. If A and B are formulas, then $(A) \supset (B)$ is a *formula*.
3. If A and B are formulas, then $(A) \,\&\, (B)$ is a *formula*.
4. If A and B are formulas, then $(A) \vee (B)$ is a *formula*.
5. If A is a formula, then $\sim (A)$ is a *formula*.
6. If x is a variable and A is a formula, then $\forall_x (A)$ and $\exists_x (A)$ are *formulas*.
7. The only *formulas* are those given by 1–6.

THEOREM:
$(\exists_c ((((c)') + (a)) = (b))) \supset (\sim ((a) = (b)))$ is a formula.
PROOF: By the terms obtained and by 1, $(a) = (b)$ and $(((c)') + (a)) = (b)$ are formulas.
     By 5 $\sim ((a) = (b))$ is a formula.
     By 7 $\exists_c ((((c)') + (a)) = (b))$ is a formula.
     By 2 $(\exists_c ((((c)') + (a)) = (b))) \supset (\sim ((a) = (b)))$ is a formula.

These two simple examples should indicate the procedure by which meta-mathematics proceeds to "construct" the formulas of classical mathematics.

I should like to conclude by calling your attention to what Hilbert hoped to do and how. We have already seen how the fundamental problem is one of knowing that we cannot ever reach a contradiction in the processes utilized in arithmetic.

(This, I emphasize again, is a consequence of our accepting the view that mathematics may be of reality but is not about reality.) Hilbert hoped to do this by showing that in formal mathematics we utilize certain symbols. Let us list all of them. In formal mathematics, we group them together. Let us explicitly—in our meta-mathematics—state how symbols can be grouped together. In formal mathematics, we use such concepts as substitution and deduction. Let us explicitly—in our meta-mathematics—indicate what is a permissible substitution and deduction. In formal mathematics we state what a contradiction is. Let us explicitly—in our meta-mathematics—state what combinations of symbols are contradictions. In our formal mathematical system we would like to believe that no contradictions can arise. Let us explicitly—in our meta-mathematics—prove, with finite means, that with the explicitly given symbols, and the explicitly stated formulas, and the explicitly formalized logic, it is not possible to obtain a formula which is a combination of symbols that we have called a contradiction in our meta-mathematics.

The real problem then is to be able to prove that we cannot arrive at a combination of symbols called a contradiction. The machinery to do this is meta-mathematics. It is the machinery by means of which we analyze the possible consequences of our moves with symbols. Hilbert did not succeed; that is, he was unable to prove that contradictions—certain combinations of symbols—cannot occur. At least he was unable to do this for arithmetic, i.e., the natural numbers. Such a proof of freedom from contradiction of number theory has still not been given. There is some evidence—the theorems of Kurt Gödel, for example—that it may be impossible. This may not be necessarily so. As Professor Kleene has remarked, all this may do is "pose a challenge to the meta-mathematician to bring to bear methods of finitary proof more powerful than those commonly used in elementary number theory." [14]

Meta-mathematics, then—the science created to subject classical mathematics to formal criticism—has not succeeded in giving us an absolute proof of noncontradiction. But it has, nevertheless, put in our hands powerful tools for analyzing the formal structures which we call mathematics. It has enabled us at the very least to look hard at our symbols and the operations on them with no regard to the meanings which informal mathematics puts upon them. If we really believe that modern mathematics is an abstract science, then we cannot avoid the consequences. And these consequences are precisely that we must then subject to critical analysis what we are, in fact, doing when we say we are engaged in mathematics—and this is what meta-mathematics aims to do.

NOTES

1. Delivered, with slight revisions, on March 29, 1957, at the Annual Meeting of the National Council of Teachers of Mathematics, Philadelphia.

2. For the discussion immediately following of the historical development of abstractness in algebra, I am indebted to the article by Ernest Nagel, "'Impossible Numbers': A Chapter in the History of Modern Logic," *Studies in the History of Ideas* (New York: Columbia University Press, 1935), II, 425–475.

3. *Ibid.*, 436, quoted by Nagel.

4. Duncan F. Gregory, "On the Real Nature of Symbolical Algebra," *Edinburgh Philosophical Transactions*, XIV, Part I (1838), 208.

5. A. De Morgan, "On the Foundations of Algebra," *Transactions of the Cambridge Philosophical Society*, VII (1839), 173.

6. See D. Hilbert, *The Foundations of Geometry* (trans. by E. J. Townsend) (La Salle, Illinois: Open Court Publishing Co., 1950, reprint of 1902 edition), pp. 25–36.

7. D. Hilbert, "On the Foundations of Logic and Arithmetic" (trans. by G. B. Halstead), *The Monist*, XV (1905), 338–352.

8. *Ibid.*, 340.

9. See for example, A. Church, "The Richard Paradox," *American Mathematical Monthly*, 41 (1934), 356–361.

10. It should be noted that Hilbert abandoned his early enthusiasm for the point of view of the Logistic School (Frege-Russell).

11. This argument is especially compelling when dealing with infinite systems when sense intuition often abandons us completely.

12. The following discussion is taken from J. Von Neumann, "Die formalistische Grundlegung der Mathematik," *Erkenntnis*, II (1931), 116 ff.

13. The following discussion is indebted to S. C. Kleene, *Introduction to Meta-Mathematics* (New York: D. Van Nostrand Co., Inc., 1952), pp. 72 ff.

14. *Ibid.*, 211–212.

18

Intuition and Logic in Mathematics

HENRI POINCARÉ

I

It is impossible to study the works of the great mathematicians, or even those of the lesser, without noticing and distinguishing two opposite tendencies, or rather two entirely different kinds of minds. The one sort are above all preoccupied with logic; to read their works, one is tempted to believe they have advanced only step by step, after the manner of a Vauban who pushes on his trenches against the place besieged, leaving nothing to chance. The other sort are guided by intuition and at the first stroke make quick but sometimes precarious conquests, like bold cavalrymen of the advance guard.

The method is not imposed by the matter treated. Though one often says of the first that they are *analysts* and calls the others *geometers,* that does not prevent the one sort from remaining analysts even when they work at geometry, while the others are still geometers even when they occupy themselves with pure analysis. It is the very nature of their mind which makes them logicians or intuitionalists, and they cannot lay it aside when they approach a new subject.

Nor is it education which has developed in them one of the two tendencies and stifled the other. The mathematician is born, not made, and

Reprinted from *Mathematics Teacher* 62: (Mar., 1969): 205–12; with permission of the National Council of Teachers of Mathematics. Originally published in Henri Poincaré, *The Foundations of Science,* trans. George Bruce Halsted ("Science and Education," I [New York and Lancaster: The Science Press, 1929]), 210–22.

it seems he is born a geometer or an analyst. I should like to cite examples, and there are surely plenty; but to accentuate the contrast I shall begin with an extreme example, taking the liberty of seeking it in two living mathematicians.

M. Méray wants to prove that a binominal equation always has a root, or, in ordinary words, that an angle may always be subdivided. If there is any truth that we think we know by direct intuition, it is this. Who could doubt that an angle may always be divided into any number of equal parts? M. Méray does not look at it that way; in his eyes this proposition is not at all evident and to prove it he needs several pages.

On the other hand, look at Professor Klein: he is studying one of the most abstract questions of the theory of functions: to determine whether on a given Riemann surface there always exists a function admitting of given singularities. What does the celebrated German geometer do? He replaces his Riemann surface by a metallic surface whose electric conductivity varies according to certain laws. He connects two of its points with the two poles of a battery. The current, says he, must pass, and the distribution of this current on the surface will define a function whose singularities will be precisely those called for by the enunciation.

Doubtless Professor Klein well knows he has given here only a sketch; nevertheless he has not hesitated to publish it; and he would probably believe he finds in it, if not a rigorous demonstration, at least a kind of moral certainty. A logician would have rejected with horror such a conception, or rather he would not have had to reject it, because in his mind it would never have originated.

Again, permit me to compare two men, the honor of French science, who have recently been taken from us, but who both entered long ago into immortality. I speak of M. Bertrand and M. Hermite. They were scholars of the same school at the same time; they had the same education, were under the same influences; and yet what a difference! Not only does it blaze forth in their writings; it is in their teaching, in their way of speaking, in their very look. In the memory of all their pupils these two faces are stamped in deathless lines; for all who have had the pleasure of following their teaching, this remembrance is still fresh; it is easy for us to evoke it.

While speaking, M. Bertrand is always in motion; now he seems in combat with some outside enemy, now he outlines with a gesture of the hand the figures he studies. Plainly he sees and he is eager to paint, this is why he calls gesture to his aid. With M. Hermite, it is just the opposite, his eyes seem to shun contact with the world; it is not without, it is within he seeks the vision of truth.

Among the German geometers of this century, two names above all are illustrious, those of the two scientists who founded the general theory of functions, Weierstrass and Riemann. Weierstrass leads everything back to the consideration of series and their analytic transformations; to express it better, he reduces analysis to a sort of prolongation of arithmetic; you may turn through all his books without finding a figure. Riemann, on the contrary, at once calls geometry to his aid; each of his conceptions is an image that no one can forget, once he has caught its meaning. . . .

Among our students we notice the same differences; some prefer to treat their problems "by analysis," others "by geometry." The first are incapable of "seeing in space," the others are quickly tired of long calculations and become perplexed.

The two sorts of minds are equally necessary for the progress of science; both the logicians and the intuitionalists have achieved great things that others could not have done. Who would venture to say whether he preferred that Weierstrass had never written or that there had never been a Riemann? Analysis and synthesis have then both their legitimate roles. But it is interesting to study more closely in the history of science the part which belongs to each.

II

Strange! If we read over the works of the ancients we are tempted to class them all among the intuitionalists. And yet nature is always the same; it is hardly probable that it has begun in this century to create minds devoted to logic. If we could put ourselves into the flow of ideas which reigned in their time, we should recognize that many of the old geometers were in tendency analysts. Euclid, for example, erected a scientific structure wherein his contemporaries could find no fault. In this vast construction, of which each piece however is due to intuition, we may still today, without much effort, recognize the work of a logician.

It is not minds that have changed, it is ideas; the intuitional minds have remained the same; but their readers have required of them greater concessions.

What is the cause of this evolution? It is not hard to find. Intuition can not give us rigor, nor even certainty; this has been recognized more and more. Let us cite some examples. We know there exist continuous functions lacking derivatives. Nothing is more shocking to intuition than this proposition which is imposed upon us by logic. Our fathers would not have failed to say: "It is evident that every continuous function has a derivative, since every curve has a tangent."

How can intuition deceive us on this point? It is because when we seek to imagine a curve we can not represent it to ourselves without width; just so, when we represent to ourselves a straight line, we see it under the form of a rectilinear band of a certain breadth. We well know these lines have no width; we try to imagine them narrower and narrower and thus to approach the limit; so we do in a certain measure, but we shall never attain this limit. And then it is clear we can always picture these two narrow bands, one straight, one curved, in a position such that they encroach slightly one upon the other without crossing. We shall thus be led, unless warned by a rigorous analysis, to conclude that a curve always has a tangent.

I shall take as second example Dirichlet's principle on which rest so many theorems of mathematical physics; today we establish it by reasoning very rigorous but very long; heretofore, on the contrary, we were content with a very summary proof. A certain integral depending on an arbitrary function can never vanish. Hence it is concluded that it must have a minimum. The flaw in this reasoning strikes us immediately, since we use the abstract term *function* and are familiar with all the singularities functions can present when the word is understood in the most general sense.

But it would not be the same had we used concrete images, had we, for example, considered this function as an electric potential; it would have been thought legitimate to affirm that electrostatic equilibrium can be attained. Yet perhaps a physical comparison would have awakened some vague distrust. But if care had been taken to translate the reasoning into the language of geometry, intermediate between that of analysis and that of physics, doubtless this distrust would not have been produced, and perhaps one might thus, even today, still deceive many readers not forewarned.

Intuition, therefore, does not give us certainty. This is why the evolution had to happen; let us now see how it happened.

It was not slow in being noticed that rigor could not be introduced in the reasoning unless first made to enter into the definitions. For the most part the objects treated of by mathematicians were long ill-defined; they were supposed to be known because represented by means of the senses or the imagination; but one had only a crude image of them and not a precise idea on which reasoning could take hold. It was there first that the logicians had to direct their efforts.

So, in the case of incommensurable numbers. The vague idea of continuity, which we owe to intuition, resolved itself into a complicated system of inequalities referring to whole numbers.

By that means the difficulties arising from passing to the limit, or from the consideration of infinitesimals, are finally removed. Today in analysis only whole numbers are left or systems, finite or infinite, of whole numbers bound together by a net of equality or inequality relations. Mathematics, as they say, is arithmetized.

III

A first question presents itself. Is this evolution ended? Have we finally attained absolute rigor? At each stage of the evolution our fathers also thought they had reached it. If they deceived themselves, do we not likewise cheat ourselves?

We believe that in our reasonings we no longer appeal to intuition; the philosophers will tell us this is an illusion. Pure logic could never lead us to anything but tautologies; it could create nothing new; not from it alone can any science issue. In one sense these philosophers are right; to make arithmetic, as to make geometry, or to make any science, something else than pure logic is necessary. To designate this something else we have no word other than *intuition*. But how many different ideas are hidden under this same word?

Compare these four axioms: (1) Two quantities equal to a third are equal to one another; (2) if a theorem is true of the number 1 and if we prove that it is true of $n + 1$ if true for n, then will it be true of all whole numbers; (3) if on a straight the point C is between A and B and the point D between A and C, then the point D will be between A and B; (4) through a given point there is not more than one parallel to a given straight.

All four are attributed to intuition, and yet the first is the enunciation of one of the rules of formal logic; the second is a real synthetic a priori judgment, it is the foundation of rigorous mathematical induction; the third is an appeal to the imagination; the fourth is a disguised definition.

Intuition is not necessarily founded on the evidence of the senses; the senses would soon become powerless; for example, we can not represent to ourselves a chiliagon, and yet we reason by intuition on polygons in general, which include the chiliagon as a particular case.

You know what Poncelet understood by the *principle of continuity*. What is true of a real quantity, said Poncelet, should be true of an imaginary quantity; what is true of the hyperbola whose asymptotes are real should then be true of the ellipse whose asymptotes are imaginary. Poncelet was one of the most intuitive minds of this century; he was passionately, almost ostentatiously, so; he regarded the principle of continuity as one of his boldest conceptions, and yet this principle did not rest on the evidence of the senses. To assimilate the hyperbola to the ellipse was rather to contradict this evidence. It was only a sort of precocious and instinctive generalization which, moreover, I have no desire to defend.

We have then many kinds of intuition: first, the appeal to the senses and the imagination; next, generalization by induction, copied, so to speak, from the procedures of the experimental sciences; finally, we have the intuition of pure number, whence arose the second of the axioms just enunciated, which is able to create the real mathematical reasoning. I have shown above by examples that the first two can not give us certainty; but who will seriously doubt the third, who will doubt arithmetic?

Now in the analysis of today, when one cares to take the trouble to be rigorous, there can be nothing but syllogisms or appeals to this intuition of pure number, the only intuition which can not deceive us. It may be said that today absolute rigor is attained.

IV

The philosophers make still another objection: "What you gain in rigor," they say, "you lose in objectivity. You can rise toward your logical ideal only by cutting the bonds which attach you to reality. Your science is infallible, but it can only remain so by imprisoning itself in an ivory tower and renouncing all relation with the external world. From this seclusion it must go out when it would attempt the slightest application."

For example, I seek to show that some property pertains to some object whose concept seems to me at first indefinable, because it is intuitive. At first I fail or must content myself with approximate proofs; finally I decide to give to my object a precise definition, and this enables me to establish this property in an irreproachable manner.

"And then," say the philosophers, "it still remains to show that the object which corresponds to this definition is indeed the same made known to you by intuition; or else that some real and concrete object whose conformity with your intuitive idea you believe you immediately recognize corresponds to your new definition. Only then could you affirm that it has the property in question. You have only displaced the difficulty."

That is not exactly so; the difficulty has not been displaced, it has been divided. The proposition to be established was in reality composed of two different truths, at first not distinguished. The first was a mathematical truth, and it is now rigorously established. The second was an experimental verity. Experience alone can teach us that some real and concrete object corresponds or does not correspond to some abstract definition. This second verity is not mathematically demonstrated, but neither can it be, any more than can the empirical laws of the physical and natural sciences. It would be unreasonable to ask more.

Well, is it not a great advance to have distinguished what long was wrongly confused? Does this mean that nothing is left of this objection of the philosophers? That I do not intend to say; in becoming rigorous, mathematical science takes a character so artificial as to strike every one; it forgets its historical origins; we see how the questions can be answered, we no longer see how and why they are put.

This shows us that logic is not enough; that the science of demonstration is not all science and that intuition must retain its role as complement, I was about to say as counterpoise or as antidote of logic.

I have already had occasion to insist on the place intuition should hold in the teaching of the mathematical sciences. Without it young minds could not make a beginning in the understanding of mathematics; they could not learn to love it and would see in it only a vain logomachy; above all, without intuition they would never become capable of applying mathematics. But now I wish before all to speak of the role of intuition in science itself. If it is useful to the student it is still more so to the creative scientist.

V

We seek reality, but what is reality? The physiologists tell us that organisms are formed of cells; the chemists add that cells themselves are formed of atoms. Does this mean that these atoms or these cells constitute reality, or rather the sole reality? The way in which these cells are arranged and from which results the unity of the individual, is not it also a reality much more interesting than that of the isolated elements, and should a naturalist who had never studied the elephant except by means of the microscope think himself sufficiently acquainted with that animal?

Well, there is something analogous to this in mathematics. The logician cuts up, so to speak, each demonstration into a very great number of elementary operations; when we have examined these operations one after the other and ascertained that each is correct, are we to think we have grasped the real meaning of the demonstration? Shall we have understood it even when, by an effort of memory, we have become able to repeat this proof by reproducing all these elementary operations in just the order in which the inventor had

arranged them? Evidently not; we shall not yet possess the entire reality; that I know not what, which makes the unity of the demonstration, will completely elude us.

Pure analysis puts at our disposal a multitude of procedures whose infallibility it guarantees; it opens to us a thousand different ways on which we can embark in all confidence; we are assured of meeting there no obstacles; but of all these ways, which will lead us most promptly to our goal? Who shall tell us which to choose? We need a faculty which makes us see the end from afar, and intuition is this faculty. It is necessary to the explorer for choosing his route; it is not less so to the one following his trail who wants to know why he chose it.

If you are present at a game of chess, it will not suffice, for the understanding of the game, to know the rules for moving the pieces. That will only enable you to recognize that each move has been made conformably to these rules, and this knowledge will truly have very little value. Yet this is what the reader of a book on mathematics would do if he were a logician only. To understand the game is wholly another matter; it is to know why the player moves this piece rather than that other which he could have moved without breaking the rules of the game. It is to perceive the inward reason which makes of this series of successive moves a sort of organized whole. This faculty is still more necessary for the player himself, that is, for the inventor.

Let us drop this comparison and return to mathematics. For example, see what has happened to the idea of continuous function. At the outset this was only a sensible image, for example, that of a continuous mark traced by the chalk on the blackboard. Then it became little by little more refined; ere long it was used to construct a complicated system of inequalities, which reproduced, so to speak, all the lines of the original image; this construction finished, the centering of the arch, so to say, was removed, that crude representation which had temporarily served as support and which was afterward useless was rejected; there remained only the construction itself, irreproachable in the eyes of the logician. And yet if the primitive image had totally disappeared from our recollection, how could we divine by what caprice all these inequalities were erected in this fashion one upon another?

Perhaps you think I use too many comparisons; yet pardon still another. You have doubtless seen those delicate assemblages of silicious needles which form the skeleton of certain sponges. When the organic matter has disappeared, there remains only a frail and elegant lace-work. True, nothing is there except silica, but what is interesting is the form this silica has taken, and we could not understand it if we did not know the living sponge which has given it precisely this form. Thus it is that the old intuitive notions of our fathers, even when we have abandoned them, still imprint their forms upon the logical constructions we have put in their place.

This view of the aggregate is necessary for the inventor; it is equally necessary for whoever wishes really to comprehend the inventor. Can logic give it to us? No; the name mathematicians give it would suffice to prove this. In mathematics logic is called *analysis* and analysis means *division, dissection*. It can have, therefore, no tool other than the scalpel and the microscope.

Thus logic and intuition have each their necessary role. Each is indispensable. Logic, which alone can give certainty, is the instrument of demonstration; intuition is the instrument of invention.

VI

But at the moment of formulating this conclusion I am seized with scruples. At the outset I distinguished two kinds of mathematical minds, the one sort logicians and analysts, the others intuitionalists and geometers. Well, the analysts also have been inventors. The names I have just cited make my insistence on this unnecessary.

Here is a contradiction, at least apparently, which needs explanation. And first, do you think

these logicians have always proceeded from the general to the particular, as the rules of formal logic would seem to require of them? Not thus could they have extended the boundaries of science; scientific conquest is to be made only by generalization.

In one of the chapters of *Science and Hypothesis*, I have had occasion to study the nature of mathematical reasoning, and have shown how this reasoning, without ceasing to be absolutely rigorous, could lead us from the particular to the general by a procedure I have called *mathematical induction*. It is by this procedure that the analysts have made science progress, and as we examine the detail itself of their demonstrations, we shall find it there at each instant beside the classic syllogism of Aristotle. We, therefore, see already that the analysts are not simply makers of syllogisms after the fashion of the scholastics.

Besides, do you think they have always marched step by step with no vision of the goal they wished to attain? They must have divined the way leading thither, and for that they needed a guide. This guide is, first, analogy. For example, one of the methods of demonstration dear to analysts is that founded on the employment of dominant functions. We know it has already served to solve a multitude of problems; in what consists then the role of the inventor who wishes to apply it to a new problem? At the outset he must recognize the analogy of this question with those which have already been solved by this method; then he must perceive in what way this new question differs from the others, and thence deduce the modifications necessary to apply to the method.

But how does one perceive these analogies and these differences? In the example just cited they are almost always evident, but I could have found others where they would have been much more deeply hidden; often a very uncommon penetration is necessary for their discovery. The analysts, not to let these hidden analogies escape them, that is, in order to be inventors, must, without the aid of the senses and imagination, have a direct sense of what constitutes the unity of a piece of reasoning, of what makes, so to speak, its soul and inmost life.

When one talked with M. Hermite, he never evoked a sensuous image, and yet you soon perceived that the most abstract entities were for him like living beings. He did not see them, but he perceived that they are not an artificial assemblage and that they have some principle of internal unity.

But, one will say, that still is intuition. Shall we conclude that the distinction made at the outset was only apparent, that there is only one sort of mind and that all the mathematicians are intuitionalists, at least those who are capable of inventing?

No, our distinction corresponds to something real. I have said above that there are many kinds of intuition. I have said how much the intuition of pure number, whence comes rigorous mathematical induction, differs from sensible intuition to which the imagination, properly so called, is the principal contributor.

Is the abyss which separates them less profound than it at first appeared? Could we recognize with a little attention that this pure intuition itself could not do without the aid of the senses? This is the affair of the psychologist and the metaphysician, and I shall not discuss the question: But the thing's being doubtful is enough to justify me in recognizing and affirming an essential difference between the two kinds of intuition; they have not the same object and seem to call into play two different faculties of our soul; one would think of two searchlights directed upon two worlds strangers to one another.

It is the intuition of pure number, that of pure logical forms, which illumines and directs those we have called *analysts*. This it is which enables them not alone to demonstrate, but also to invent. By it they perceive at a glance the general plan of a logical edifice, and that too without the senses appearing to intervene. In rejecting the aid of the imagination, which, as we have seen, is not always infallible, they can advance without fear of deceiving themselves. Happy, therefore, are those who can do without this aid! We must admire them; but how rare they are!

Among the analysts there will then be inventors, but they will be few. The majority of us, if we wished to see afar by pure intuition alone, would soon feel ourselves seized with vertigo. Our weakness has need of a staff more solid, and, despite the exceptions of which we have just spoken, it is none the less true that sensible intuition is in mathematics the most usual instrument of invention.

Apropos of these reflections, a question comes up that I have not the time either to solve or even to enunciate with the developments it would admit of. Is there room for a new distinction, for distinguishing among the analysts those who above all use pure intuition and those who are first of all preoccupied with formal logic?

M. Hermite, for example, whom I have just cited, can not be classed among the geometers who make use of the sensible intuition; but neither is he a logician, properly so called. He does not conceal his aversion to purely deductive procedures which start from the general and end in the particular.

19

The Three Crises in Mathematics: Logicism, Intuitionism, and Formalism

ERNST SNAPPER

$\mathcal{T}$HE THREE SCHOOLS, mentioned in the title, all tried to give a firm foundation to mathematics. The three crises are the failures of these schools to complete their tasks. This article looks at these crises "through modern eyes," using whatever mathematics is available today and not just the mathematics which was available to the pioneers who created these schools. Hence, this article does not approach the three crises in a strictly historical way. This article also does not discuss the large volume of current, technical mathematics which has arisen out of the techniques introduced by the three schools in question. One reason is that such a discussion would take a book and not a short article. Another one is that all this technical mathematics has very little to do with the philosophy of mathematics, and in this article I want to stress those aspects of logicism, intuitionism, and formalism which show clearly that these schools are founded in philosophy.

Logicism

This school was started in about 1884 by the German philosopher, logician, and mathematician, Gottlob Frege (1848–1925). The school was rediscovered about eighteen years later by Bertrand Russell. Other early logicists were Peano and

Russell's coauthor of *Principia Mathematica*, A. N. Whitehead. The purpose of logicism was to show that classical mathematics is part of logic. If the logicists had been able to carry out their program successfully, such questions as "Why is classical mathematics free of contradictions?" would have become "Why is logic free of contradictions?" This latter question is one on which philosophers have at least a thorough handle and one may say in general that the successful completion of the logicists' program would have given classical mathematics a firm foundation in terms of logic.

Clearly, in order to carry out this program of the logicists, one must first, somehow, define what "classical mathematics" is and what "logic" is. Otherwise, what are we supposed to show is part of what? It is precisely at these two definitions that we want to look through modern eyes, imagining that the pioneers of logicism had all of present-day mathematics available to them. We begin with classical mathematics.

In order to carry out their program, Russell and Whitehead created *Principia Mathematica* [10] which was published in 1910. (The first volume of this classic can be bought for $3.45! Thank heaven, only modern books and not the classics have become too expensive for the average reader.) *Principia*, as we will refer to *Principia Mathematica*, may be considered as a formal set theory. Although the formalization was not entirely complete, Russell and Whitehead thought that it was

Reprinted from *Mathematics Magazine* 52 (Sept., 1979): 207–16; with permission of the Mathematical Association of America.

and planned to use it to show that mathematics can be reduced to logic. They showed that all classical mathematics, known in their time, can be derived from set theory and hence from the axioms of *Principia*. Consequently, what remained to be done, was to show that all the axioms of *Principia* belong to logic.

Of course, instead of *Principia*, one can use any other formal set theory just as well. Since today the formal set theory developed by Zermelo and Fraenkel (ZF) is so much better known than *Principia*, we shall from now on refer to ZF instead of *Principia*. ZF has only nine axioms and, although several of them are actually axiom schemas, we shall refer to all of them as "axioms." The formulation of the logicists' program now becomes: Show that all nine axioms of ZF belong to logic.

This formulation of logicism is based on the thesis that classical mathematics can be defined as the set of theorems which can be proved within ZF. This definition of classical mathematics is far from perfect, as is discussed in [12]. However, the above formulation of logicism is satisfactory for the purpose of showing that this school was not able to carry out its program. We now turn to the definition of logic.

* * *

In order to understand logicism, it is very important to see clearly what the logicists meant by "logic." The reason is that, whatever they meant, they certainly meant more than classical logic. Nowadays, one can define classical logic as consisting of all those theorems which can be proven in first order languages (discussed below in the section on formalism) without the use of nonlogical axioms. We are hence restricting ourselves to first order logic and use the deduction rules and logical axioms of that logic. An example of such a theorem is the law of the excluded middle which says that, if p is a proposition, then either p or its negation $\neg p$ is true; in other words, the proposition $p \vee \neg p$ is always true where $\vee$ is the usual symbol for the inclusive "or."

If this definition of classical logic had also been the logicists' definition of logic, it would be a folly to

think for even one second that all of ZF can be reduced to logic. However, the logicists' definition was more extensive. They had a general concept as to when a proposition belongs to logic, that is, when a proposition should be called a "logical proposition." They said: *A logical proposition is a proposition which has complete generality and is true in virtue of its form rather than its content.* Here, the word "proposition" is used as synonymous with "theorem."

For example, the above law of the excluded middle "$p \vee \neg p$" is a logical proposition. Namely, this law does not hold because of any special content of the proposition p; it does not matter whether p is a proposition of mathematics or physics or what have you. On the contrary, this law holds with "complete generality," that is, for any proposition p whatsoever. Why then does it hold? The logicists answer: "Because of its form." Here they mean by form "syntactical form," the form of $p \vee \neg p$ being given by the two connectives of everyday speech, the inclusive "or" and the negation "not" (denoted by $\vee$ and $\neg$, respectively.)

On the one hand, it is not difficult to argue that all theorems of classical logic, as defined above, are logical propositions in the sense of logicism. On the other hand, there is no *a priori* reason to believe that there could not be logical propositions which lie outside of classical logic. This is why we said that the logicists' definition of logic is more extensive than the definition of classical logic. And now the logicists' task becomes clearer: It consists in showing that all nine axioms of ZF are logical propositions in the sense of logicism.

The only way to assess the success or failure of logicism in carrying out this task is by going through all nine axioms of ZF and determining for each of them whether it falls under the logicists' concept of a logical proposition. This would take a separate article and would be of interest only to readers who are thoroughly familiar with ZF. Hence, instead, we simply state that at least two of these axioms, namely, the axiom of infinity and the axiom of choice, cannot possibly be considered as logical propositions. For example, the axiom of infinity

says that there exist infinite sets. Why do we accept this axiom as being true? The reason is that everyone is familiar with so many infinite sets, say, the set of the natural numbers or the set of points in Euclidean 3-space. Hence, we accept this axiom on grounds of our everyday experience with sets, and this clearly shows that we accept it in virtue of its content and not in virtue of its syntactical form. In general, when an axiom claims the existence of objects with which we are familiar on grounds of our common everyday experience, it is pretty certain that this axiom is not a logical proposition in the sense of logicism.

And here then is the first crisis in mathematics: Since at least two out of the nine axioms of ZF are not logical propositions in the sense of logicism, it is fair to say that this school failed by about 20% in its effort to give mathematics a firm foundation. However, logicism has been of the greatest importance for the development of modern mathematical logic. In fact, it was logicism which started

mathematical logic in a serious way. The two quantifiers, the "for all" quantifier $\forall$ and the "there exists" quantifier $\exists$ were introduced into logic by Frege [5], and the influence of *Principia* on the development of mathematical logic is history.

It is important to realize that logicism is founded in philosophy. For example, when the logicists tell us what they mean by a logical proposition (above), they use philosophical and not mathematical language. They have to use philosophical language for that purpose since mathematics simply cannot handle definitions of so wide a scope.

The philosophy of logicism is sometimes said to be based on the philosophical school called "realism." In medieval philosophy "realism" stood for the Platonic doctrine that abstract entities have an existence independent of the human mind. Mathematics is, of course, full of abstract entities such as numbers, functions, sets, etc., and according to Plato all such entities exist outside our mind. The mind can discover them but does not create them. This doctrine has the advantage that one can accept such a concept as "set" without worrying about how the mind can construct a set. According to realism, sets are there for us to discover, not to be constructed, and the same holds for all other abstract entities. In short, realism allows us to accept many more abstract entities in mathematics than a philosophy which had limited us to accepting only those entities the human mind can construct. Russell was a realist and accepted the abstract entities which occur in classical mathematics without questioning whether our own minds can construct them. This is the fundamental difference between logicism and intuitionism, since in intuitionism abstract entities are admitted only if they are man made.

Excellent expositions of logicism can be found in Russell's writing, for example [9], [10], and [11].

Intuitionism

This school was begun about 1908 by the Dutch mathematician, L. E. J. Brouwer (1881–1966). The intuitionists went about the foundations of mathematics in a radically different way from the logicists. The logicists never thought that there was anything wrong with classical mathematics; they simply wanted to show that classical mathematics is part of logic. The intuitionists, on the contrary, felt that there was plenty wrong with classical mathematics.

By 1908, several paradoxes had arisen in Cantor's set theory. Here, the word "paradox" is used as synonymous with "contradiction." Georg Cantor created set theory, starting around 1870, and he did his work "naively," meaning nonaxiomatically. Consequently, he formed sets with such abandon that he himself, Russell, and others found several paradoxes within his theory. The logicists considered these paradoxes as common errors, caused by erring mathematicians and not by a faulty mathematics. The intuitionists, on the other hand, considered these paradoxes as clear indications that classical mathematics itself is far from perfect. They

felt that mathematics had to be rebuilt from the bottom on up.

The "bottom," that is, the beginning of mathematics for the intuitionists, is their explanation of what the natural numbers 1, 2, 3, … are. (Observe that we do not include the number zero among the natural numbers.) According to intuitionistic philosophy, all human beings have a primordial intuition for the natural numbers within them. This means in the first place that we have an immediate certainty as to what is meant by the number 1 and, secondly, that the mental process which goes into the formation of the number 1 can be repeated. When we do repeat it, we obtain the concept of the number 2; when we repeat it again, the concept of the number 3; in this way, human beings can construct any *finite* initial segment 1, 2, …, n for any natural number n. This mental construction of one natural number after the other would never have been possible if we did not have an awareness of time within us. "After" refers to time and Brouwer agrees with the philosopher Immanuel Kant (1724–1804) that human beings have an immediate awareness of time. Kant used the word "intuition" for "immediate awareness" and this is where the name "intuitionism" comes from. (See Chapter IV of [4] for more information about this intuitionistic concept of natural numbers.)

It is important to observe that the intuitionistic construction of natural numbers allows one to construct only arbitrarily long *finite* initial segments 1, 2, …, n. It does not allow us to construct that whole closed set of all the natural numbers which is so familiar from classical mathematics. It is equally important to observe that this construction is both "inductive" and "effective." It is inductive in the sense that, if one wants to construct, say, the number 3, one has to go through all the mental steps of first constructing the 1, then the 2, and finally the 3; one cannot just grab the number 3 out of the sky. It is effective in the sense that, once the construction of a natural number has been finished, that natural number has been constructed in its entirety. It stands before us as a completely finished mental construct, ready for our study of it. When someone says, "I have finished the mental construction of the number 3," it is like a brick-layer saying, "I have finished that wall," which he can say only after he has laid every stone in place.

We now turn to the intuitionistic definition of mathematics. According to intuitionistic philosophy, mathematics should be defined as a mental activity and not as a set of theorems (as was done above in the section on logicism). It is the activity which consists in carrying out, one after the other, those mental constructions which are inductive and effective in the sense in which the intuitionistic construction of the natural numbers is inductive and effective. Intuitionism maintains that human beings are able to recognize whether a given mental construction has these two properties. We shall refer to a mental construction which has these two properties as a *construct* and hence the intuitionistic definition of mathematics says: *Mathematics is the mental activity which consists in carrying out constructs one after the other.*

A major consequence of this definition is that all of intuitionistic mathematics is effective or "constructive" as one usually says. We shall use the adjective "constructive" as synonymous with "effective" from now on. Namely, every construct is constructive, and intuitionistic mathematics is nothing but carrying out constructs over and over. For instance, if a real number r occurs in an intuitionistic proof or theorem, it never occurs there merely on grounds of an existence proof. It occurs there because it has been constructed from top to bottom. This implies for example that each decimal place in the decimal expansion of r can in principle be computed. In short, all intuitionistic proofs, theorems, definitions, etc., are entirely constructive.

Another major consequence of the intuitionistic definition of mathematics is that mathematics cannot be reduced to any other science such as, for instance, logic. This definition comprises too many mental processes for such a reduction. And here, then, we see a radical difference between logicism and intuitionism. In fact, the intuitionistic attitude

toward logic is precisely the opposite from the logicists' attitude: According to the intuitionists, whatever valid logical processes there are, they are all constructs; hence, the valid part of classical logic is part of mathematics! Any law of classical logic which is not composed of constructs is for the intuitionist a meaningless combination of words. It was, of course, shocking that the classical law of the excluded middle turned out to be such a meaningless combination of words. This implies that this law cannot be used indiscriminately in intuitionistic mathematics; it can often be used, but not always.

* * *

Once the intuitionistic definition of mathematics has been understood and accepted, all there remains to be done is to do mathematics the intuitionistic way. Indeed, the intuitionists have developed intuitionistic arithmetic, algebra, analysis, set theory, etc. However, in each of these branches of mathematics, there occur classical theorems which are not composed of constructs and, hence, are meaningless combinations of words for the intuitionists. Consequently, one cannot say that the intuitionists have reconstructed all of classical mathematics. This does not bother the intuitionists since whatever parts of classical mathematics they cannot obtain are meaningless for them anyway. Intuitionism does not have as its purpose the justification of classical mathematics. Its purpose is to give a valid definition of mathematics and then to "wait and see" what mathematics comes out of it. Whatever classical mathematics cannot be done intuitionistically simply is not mathematics for the intuitionist. We observe here another fundamental difference between logicism and intuitionism: The logicists wanted to justify all of classical mathematics. (An excellent introduction to the actual techniques of intuitionism is [8].)

Let us now ask how successful the intuitionistic school has been in giving us a good foundation for mathematics, acceptable to the majority of mathematicians. Again, there is a sharp difference between the way this question has to be answered in the present case and in the case of logicism. Even hard-nosed logicists have to admit that their school so far has failed to give mathematics a firm foundation by about 20%. However, a hard-nosed intuitionist has every right in the world to claim that intuitionism has given mathematics an entirely satisfactory foundation. There is the meaningful definition of intuitionistic mathematics, discussed above; there is the intuitionistic philosophy which tells us why constructs can never give rise to contradictions and, hence, that intuitionistic mathematics is free of contradictions. In fact, not only this problem (of freedom from contradiction) but all other problems of a foundational nature as well receive perfectly satisfactory solutions in intuitionism.

Yet if one looks at intuitionism from the outside, namely, from the viewpoint of the classical mathematician, one has to say that intuitionism has failed to give mathematics an adequate foundation. In fact, the mathematical community has almost universally rejected intuitionism. Why has the mathematical community done this, in spite of the many very attractive features of intuitionism, some of which have just been mentioned?

One reason is that classical mathematicians flatly refuse to do away with the many beautiful theorems that are meaningless combinations of words for the intuitionists. An example is the Brouwer fixed point theorem of topology which the intuitionists reject because the fixed point cannot be constructed, but can only be shown to exist on grounds of an existence proof. This, by the way, is the same Brouwer who created intuitionism; he is equally famous for his work in (nonintuitionistic) topology.

A second reason comes from theorems which can be proven both classically and intuitionistically. It often happens that the classical proof of such a theorem is short, elegant, and devilishly clever, but not constructive. The intuitionists will of course reject such a proof and replace it by their own constructive proof of the same theorem. However, this constructive proof frequently turns out to be about ten times as long as the classical proof and

often seems, at least to the classical mathematician, to have lost all of its elegance. An example is the fundamental theorem of algebra which in classical mathematics is proved in about half a page, but takes about ten pages of proof in intuitionistic mathematics. Again, classical mathematicians refuse to believe that their clever proofs are meaningless whenever such proofs are not constructive.

Finally, there are the theorems which hold in intuitionism but are false in classical mathematics. An example is the intuitionistic theorem which says that every real-valued function which is defined for *all* real numbers is continuous. This theorem is not as strange as it sounds since it depends on the intuitionistic concept of a function: A real-valued function f is defined in intuitionism for all real numbers only if, for every real number r whose intuitionistic construction has been completed, the real number $f(r)$ can be constructed. Any obviously discontinuous function a classical mathematician may mention does not satisfy this constructive criterion. Even so, theorems such as this one seem so far out to classical mathematicians that they reject any mathematics which accepts them.

These three reasons for the rejection of intuitionism by classical mathematicians are neither rational nor scientific. Nor are they pragmatic reasons, based on a conviction that classical mathematics is better for applications to physics or other sciences than is intuitionism. They are all emotional reasons, grounded in a deep sense as to what mathematics is all about. (If one of the readers knows of a truly scientific rejection of intuitionism, the author would be grateful to hear about it.) We now have the second crisis in mathematics in front of us: It consists in the failure of the intuitionistic school to make intuitionism acceptable to at least the majority of mathematicians.

It is important to realize that, like logicism, intuitionism is rooted in philosophy. When, for instance, the intuitionists state their definition of mathematics, given earlier, they use strictly philosophical and not mathematical language. It would, in fact, be quite impossible for them to use mathematics for such a definition. The mental activity which is mathematics can be defined in philosophical terms but this definition must, by necessity, use some terms which do not belong to the activity it is trying to define.

Just as logicism is related to realism, intuitionism is related to the philosophy called "conceptualism." This is the philosophy which maintains that abstract entities exist only insofar as they are constructed by the human mind. This is very much the attitude of intuitionism which holds that the abstract entities which occur in mathematics, whether sequences or order-relations or what have you, are all mental constructions. This is precisely why one does not find in intuitionism the staggering collection of abstract entities which occur in classical mathematics and hence in logicism. The contrast between logicism and intuitionism is very similar to the contrast between realism and conceptualism.

A very good way to get into intuitionism is by studying [8], Chapter IV of [4], [2], and [13], in this order.

Formalism

This school was created in about 1910 by the German mathematician David Hilbert (1862–1943). True, one might say that there were already formalists in the nineteenth century since Frege argued against them in the second volume of his *Grundgesetze der Arithmetik* (see the book by Geach and Black under [5], pages 182–233); the first volume of the *Grundgesetze* appeared in 1893 and the second one in 1903. Nevertheless, the modern concept of formalism, which includes finitary reasoning, must be credited to Hilbert. Since modern books and courses in mathematical logic usually deal with formalism, this school is much better known today than either logicism or intuitionism. We will hence discuss only the highlights of formalism and begin by asking, "What is it that we formalize when we formalize something?"

The answer is that we formalize some given *axiomatized* theory. One should guard against confusing axiomatization and formalization. Euclid *axiomatized* geometry in about 300 B.C., but formalization started only about 2200 years later with the logicists and formalists. Examples of axiomatized theories are Euclidean plane geometry with the usual Euclidean axioms, arithmetic with the Peano axioms, ZF with its nine axioms, etc. The next question is: "How do we formalize a given axiomatized theory?"

Suppose then that some axiomatized theory T is given. Restricting ourselves to first order logic, "to formalize T" means to choose an appropriate first order language for T. The vocabulary of a first order language consists of five items, four of which are always the same and are not dependent on the given theory T. These four items are the following: (1) A list of denumerably many variables—who can talk about mathematics without using variables? (2) Symbols for the connectives of everyday speech, say $\neg$ for "not," $\wedge$ for "and," $\vee$ for the inclusive "or," $\rightarrow$ for "if then," and $\leftrightarrow$ for "if and only if"—who can talk about anything at all without using connectives? (3) The equality sign $=$; again, no one can talk about mathematics without using this sign. (4) The two quantifiers, the "for all" quantifier $\forall$ and the "there exist" quantifier $\exists$; the first one is used to say such things as "*all* complex numbers have a square root," the second one to say things like "*there exist* irrational numbers." One can do without some of the above symbols, but there is no reason to go into that. Instead, we turn to the fifth item.

Since T is an axiomatized theory, it has so called "undefined terms." One has to choose an appropriate symbol for every undefined term of T and these symbols make up the fifth item. For instance, among the undefined terms of plane Euclidean geometry, occur "point," "line," and "incidence," and for each one of them an appropriate symbol must be entered into the vocabulary of the first order language. Among the undefined terms of arithmetic occur "zero," "addition," and "multiplication," and the symbols one chooses for them are of course 0, +, and ×, respectively. The easiest theory of all to formalize is ZF

since this theory has only one undefined term, namely, the membership relation. One chooses, of course, the usual symbol ∈ for that relation. These symbols, one for each undefined term of the axiomatized theory T, are often called the "parameters" of the first order language and hence the parameters make up the fifth item.

Since the parameters are the only symbols in the vocabulary of a first order language which depend on the given axiomatized theory T, one formalizes T simply by choosing these parameters. Once this choice has been made, the whole theory T has been completely formalized. One can now express in the resulting first order language L not only all axioms, definitions, and theorems of T, but more! One can also express in L all axioms of classical logic and, consequently, also all proofs

one uses to prove theorems of T. In short, one can now proceed entirely within L, that is, entirely "formally."

But now a third question presents itself: "Why in the world would anyone want to formalize a given axiomatized theory?" After all, Euclid never saw a need to formalize his axiomatized geometry. It is important to ask this question, since even the great Peano had mistaken ideas about the real purpose of formalization. He published one of his most important discoveries in differential equations in a formalized language (very similar to a first order language) with the result that nobody read it until some charitable soul translated the article into common German.

Let us now try to answer the third question. If mathematicians do technical research in a certain branch of mathematics, say, plane Euclidean geometry, they are interested in discovering and proving the important theorems of the branch of mathematics. For that kind of technical work, formalization is usually not only no help but a definite hindrance. If, however, one asks such foundational questions as, for instance, "Why is this branch of mathematics free of contradictions?", then formalization is not just a help but an absolute necessity.

It was really Hilbert's stroke of genius to understand that formalization is the proper technique to tackle such foundational questions. What he taught us can be put roughly as follows. Suppose that T is an axiomatized theory which has been formalized in terms of the first order language L. This language has such a precise syntax that it itself can be studied as a *mathematical* object. One can ask for instance: "Can one possibly run into contradictions if one proceeds entirely formally within L, using only the axioms of T and those of classical logic, all of which have been expressed in L?" If one can prove mathematically that the answer to this question is "no," one has there a mathematical proof that the theory T is free of contradictions!

This is basically what the famous "Hilbert program" was all about. The idea was to formalize the various branches of mathematics and then to prove *mathematically* that each one of them is free of contradictions. In fact if, by means of this technique, the formalists could have just shown that ZF is free of contradictions, they would thereby already have shown that all of classical mathematics is free of contradictions, since classical mathematics can be done axiomatically in terms of the nine axioms of ZF. In short, the formalists tried to create a mathematical technique by means of which one could prove that mathematics is free of contradictions. This was the original purpose of formalism.

* * *

It is interesting to observe that both logicists and formalists formalized the various branches of mathematics, but for entirely different reasons.

The logicists wanted to use such a formalization to show that the branch of mathematics in question belongs to logic; the formalists wanted to use it to prove mathematically that that branch is free of contradictions. Since both schools "formalized," they are sometimes confused.

Did the formalists complete their program successfully? No! In 1931, Kurt Gödel showed in [6] that formalization cannot be considered as a mathematical technique by means of which one can prove that mathematics is free of contradictions. The theorem in that paper which rang the death bell for the Hilbert program concerns axiomatized theories which are free of contradictions and whose axioms are strong enough so that arithmetic can be done in terms of them. Examples of theories whose axioms are that strong are, of course, Peano arithmetic and ZF. Suppose now that T is such a theory and that T has been formalized by means of the first order language L. Then Gödel's theorem says, in nontechnical language, "No sentence of L which can be interpreted as asserting that T is free of contradictions can be proven formally within the language L." Although the interpretation of this theorem is somewhat controversial, most mathematicians have concluded from it that the Hilbert program cannot be carried out: Mathematics is not able to prove its own freedom of contradictions. Here, then, is the third crisis in mathematics.

Of course, the tremendous importance of the formalist school for present-day mathematics is well known. It was in this school that modern mathematical logic and its various offshoots, such as model theory, recursive function theory, etc., really came into bloom.

Formalism, as logicism and intuitionism, is founded in philosophy, but the philosophical roots of formalism are somewhat more hidden than those of the other two schools. One can find them, though, by reflecting a little on the Hilbert program.

Let again T be an axiomatized theory which has been formalized in terms of the first order language L. In carrying out Hilbert's program, one has to talk about the language L as one object, and while doing this, one is not talking within that safe language L itself. On the contrary, one is talking about L in ordinary, everyday language, be it English or French or what have you. While using our natural language and not the formal language L, there is of course every danger that contradictions, in fact, any kind of error, may slip in. Hilbert said that the way to avoid this danger is by making absolutely certain that, while one is talking in one's natural language about L, one uses only reasonings which are absolutely safe and beyond any kind of suspicion. He called such reasonings "finitary reasonings," but had, of course, to give a definition of them. The most explicit definition of finitary reasoning known to the author was given by the French formalist Herbrand ([7], the footnote on page 622). It says, if we replace "intuitionistic" by "finitary":

> By a finitary argument we understand an argument satisfying the following conditions: In it we never consider anything but a given finite number of objects and of functions; these functions are well defined, their definition allowing the computation of their values in a univocal way; we never state that an object exists without giving the means of constructing it; we never consider the totality of all the objects x of an infinite collection; and when we say that an argument (or a theorem) is true for all these x, we mean that, for each x taken by itself, it is possible to repeat the general argument in question, which should be considered to be merely the prototype of these particular arguments.

Observe that this definition uses philosophical and not mathematical language. Even so, no one can claim to understand the Hilbert program without an understanding of what finitary reasoning amounts to. The philosophical roots of formalism come out into the open when the formalists define what they mean by finitary reasoning.

We have already compared logicism with realism, and intuitionism with conceptualism. The philosophy which is closest to formalism is "nominalism." This is the philosophy which claims that abstract entities have no existence of any kind, neither outside the human mind as maintained by

realism, nor as mental constructions within the human mind as maintained by conceptualism. For nominalism, abstract entities are mere vocal utterances or written lines, mere names. This is where the word "nominalism" comes from, since in Latin *nominalis* means "belonging to a name." Similarly, when formalists try to prove that a certain axiomatized theory T is free of contradictions, they do not study the abstract entities which occur in T but, instead, study that first order language L which was used to formalize T. That is, they study how one can form sentences in L by the proper use of the vocabulary of L; how certain of these sentences can be proven by the proper use of those special sentences of L which were singled out as axioms; and, in particular, they try to show that no sentence of L can be proven and disproven at the same time, since they would thereby have established that the original theory T is free of contradictions. The important point is that this whole study of L is a strictly syntactical study, since no meanings or abstract entities are associated with the sentences of L. This language is investigated by considering the sentences of L as meaningless expressions which are manipulated according to explicit, syntactical rules, just as the pieces of a chess game are meaningless figures which are pushed around according to the rules of the game. For the strict formalist "to do mathematics" is "to manipulate the meaningless symbols of a first order language according to explicit, syntactical rules." Hence, the strict formalist does not work with abstract entities, such as infinite series or cardinals, but only with their meaningless names which are the appropriate expressions in a first order language. Both formalists and nominalists avoid the direct use of abstract entities, and this is why formalism should be compared with nominalism.

The fact that logicism, intuitionism, and formalism correspond to realism, conceptualism, and nominalism, respectively, was brought to light in Quine's article, "On What There Is"([1], pages 183–196). Formalism can be learned from any modern book in mathematical logic, for instance [3].

Epilogue

Where do the three crises in mathematics leave us? They leave us without a firm foundation for mathematics. After Gödel's paper [6] appeared in 1931, mathematicians on the whole threw up their hands in frustration and turned away from the philosophy of mathematics. Nevertheless, the influence of the three schools discussed in this article has remained strong, since they have given us much new and beautiful mathematics. This mathematics concerns mainly set theory, intuitionism and its various constructivist modifications, and mathematical logic with its many offshoots. However, although this kind of mathematics is often referred to as "foundations of mathematics," one cannot claim to be advancing the philosophy of mathematics just because one is working in one of these areas. Modern mathematical logic, set theory, and intuitionism with its modifications are nowadays technical branches of mathematics, just as algebra or analysis, and unless we return directly to the philosophy of mathematics, we cannot expect to find a firm foundation for our science. It is evident that such a foundation is not necessary for technical mathematical research, but there are still those among us who yearn for it. The author believes that the key to the foundations of mathematics lies hidden somewhere among the philosophical roots of logicism, intuitionism, and formalism and this is why he has uncovered these roots, three times over.

Excellent literature on the foundations of mathematics is contained in [1] and [7].

REFERENCES

[1] P. Benacerraf and H. Putnam, *Philosophy of Mathematics*, Prentice-Hall, 1964.

[2] M. Dummett, *Elements of Intuitionism*, Clarendon Press, Oxford, England, 1977.

[3] H. B. Enderton, *A Mathematical Introduction to Logic*, Academic Press, 1972.

[4] A. A. Fraenkel, Y. Bar-Hillel, and A. Levy, *Foundations of Set Theory*, North-Holland, Amsterdam, Netherlands, 1973.

[5] G. Frege, *Begriffschrift*, in *Translations from the Philosophical Writings of Gottlob Frege* by P. Geach and M. Black, Basil Blackwell, Oxford, England, 1970. Also in [7] pp. 1–82.

[6] K. Gödel, On formally undecidable propositions of *Principia Mathematica* and related systems, in [7] pp. 596–616.

[7] J. van Heijenoort, *From Frege to Gödel*, Harvard Univ. Press, Cambridge. Available in paperback.

[8] A. Heyting, *Intuitionism, An Introduction*, North-Holland, Amsterdam, Netherlands, 1966.

[9] B. Russell, *Principles of Mathematics*, 1st ed. (1903) W. W. Norton, New York. Available in paperback.

[10] B. Russell and A. N. Whitehead, *Principia Mathematica*, 1st ed. (1910) Cambridge Univ. Press, Cambridge, England. Available in paperback.

*[11] B. Russell, *Introduction to Mathematical Philosophy*, Simon and Schuster, New York, 1920. Available in paperback.

[12] E. Snapper, What is mathematics?, *Amer. Math. Monthly*, no. 7, 86 (1979) 551–557.

[13] A. S. Troelstra, *Choice Sequences*, Oxford Univ. Press, Oxford, England, 1977.

*Available as a Dover reprint.

20

Kurt Gödel,
Mathematician and Logician

GERALD E. LENZ

O<small>N 15 JANUARY</small> 1978 the *New York Times* reported the death of Kurt Gödel, mathematician and logician. On 20 January, *Time* magazine carried a brief summary of Gödel's work. It is not often that the demise of a mathematician receives attention from such popular publications.

Who was Kurt Gödel and what were his contributions to mathematics? Why should he have been described (*New York Times*, 22 January 1978) as the "discoverer of the most significant mathematical truth of this century, incomprehensible to laymen, revolutionary for philosophers and logicians?"

Born in what is now Czechoslovakia in 1906, Gödel received his Ph.D. from the University of Vienna in 1930 and became a naturalized U.S. citizen in 1948. His most famous contribution to the foundations of mathematics was made when he was only twenty-five years old, in 1931, while still at the University of Vienna.

His work at that time involved the much sought after proof that there can never be any contradictions in mathematics. A demonstration of the consistency of mathematics (i.e., the impossibility of proving contradictory statements if the rules or axioms are followed) was of growing importance because mathematics had moved further away from the concrete world of experience in the late nineteenth and early

Kurt Gödel
Sketch courtesy of Christopher Munoz

twentieth centuries. The existence of hyperbolic geometry, as demonstrated by Gauss, Lobachevski, and Bolyai, presented the mathematics community with a geometric system that could not easily be reconciled with the more intuitive Euclidean geometry. This shift from the mathematics of the familiar (where consistency was not questioned) to more abstract kinds of mathematics motivated the search for consistency proofs. The German mathematician David Hilbert was one of the leading advocates of such proofs. Demonstrations of what can be proved about the formal language of mathematics are termed metamathematical.

Gödel's amazing discovery was that no such proof can ever be constructed. His results are based on the development of mathematics given by Bertrand Russell and Alfred North Whitehead in *Principia Mathematica*, but they also hold for related systems.

Reprinted from *Mathematics Teacher* 73 (Nov., 1980): 612–14; with permission of the National Council of Teachers of Mathematics.

What Gödel showed was that no axiom system for mathematics as we know it is powerful enough to lead to a proof of its own consistency. Such a proof would require that additional axioms be added to the system. The consistency of this new, larger set of axioms would then be in doubt.

Gödel's 1931 results go further. In his 1931 paper, entitled *On Formally Undecidable Propositions of Principia Mathematica and Related Systems I*, he demonstrated that if mathematics is consistent, then it is not complete. That is, if there are no contradictions in mathematics, then there exist mathematical statements that can be neither proved nor disproved. Such statements are said to be formally undecidable.

Thus at the age of twenty-five Kurt Gödel showed that the axiomatic method in mathematics has a fatal weakness. His work is important not only because of what it says about the foundations of mathematics but also because of the philosophical significance of his results for the analysis of all knowledge.

Of course there are many formally *decidable* propositions in mathematics. The task of determining which propositions are formally decidable within a given system is important in current research in mathematical logic.

On 20 September 1938 Kurt Gödel married Adele Porkert. Gödel, who had been a visiting member of the Institute for Advanced Study in Princeton, New Jersey, as early as 1933, became a permanent member of the Institute in 1946. His application for U.S. citizenship was supported by Albert Einstein, who had been Gödel's colleague at the Institute. Einstein had received a lifetime appointment to the Institute in 1933, the year he came to the U.S. to escape Hitler's Germany. Throughout the years Gödel received many awards and honorary degrees, including the United States's highest recognition in science, the National Medal of Science, presented by Gerald Ford in 1975.

One of Gödel's greatest achievements occurred, however, in 1939. We shall quickly review some of the information necessary to understand those results.

Two sets are in one-to-one correspondence if there is some way to match all the elements of both sets such that each element in one is matched to one and only one element from the other. Any two finite sets having the same number of elements are in one-to-one correspondence. However, for infinite sets the situation is more interesting.

For example, the set of all integers is in one-to-one correspondence with the set of even integers, and the set of integers is also in one-to-one correspondence with the rational numbers. However, the integers are not in one-to-one correspondence with the real numbers. In a sense the reals constitute a set that is "too big" to be in one-to-one correspondence with the integers. However, the reals are in one-to-one correspondence with all the points on a Euclidean circle and with all the complex numbers.

This approach to the comparison of infinite sets is due to Georg Cantor, who published it in the late nineteenth century. It was Cantor himself who raised the question concerning the existence of a set that is "too large" to be in one-to-one correspondence with the integers but "too small" to be in one-to-one correspondence with the reals. The assertion that there is no such set became known as the continuum hypothesis. In 1900 Hilbert placed the proof of the continuum hypothesis first in a list of the most important problems facing mathematicians in the twentieth century.

The problem proved to be difficult indeed. It was not until Gödel's 1939 work that even a partial solution was available. Gödel demonstrated that the continuum hypothesis is consistent with the other axioms of set theory. In other words, it is impossible to disprove Cantor's hypothesis. This remarkable result is reminiscent of the situation of the parallel axiom in Euclidean geometry. It can be demonstrated that the parallel axiom cannot be disproved using the other axioms for geometry. However, in Euclidean geometry it is also known that the parallel axiom cannot be proved as a theorem.

Whether or not the continuum hypothesis could be proved remained an open question until 1963. In that year Paul Cohen, who had been a fellow at the Institute for Advanced Study at Princeton from 1959 to 1961, proved that the continuum hypothesis is independent of the other axioms of set theory. In other words it cannot be proved as a theorem using only the other axioms. Thus the status of the continuum hypothesis in set theory is analogous to that of the parallel axiom in geometry.

Gödel is responsible for many other discoveries. In 1939 he showed that the axiom of choice is also independent of the other axioms of set theory. (The axiom of choice says that if one has an infinite collection of sets, it is always possible to form a new set consisting of one element from each of the given sets.) Earlier, Gödel had shown that first-order logic is complete. He also worked in physics, notably with Einstein's field equation. His work there led to the "rotating universe" model, which allows (at least in theory) for the possibility of an agent influencing the past.

The man who accomplished all this was somewhat of a private person despite the fact that he had honorary degrees from Princeton, Yale, and Harvard and was the recipient of the first Albert Einstein Award for achievement.

Kurt Gödel died of heart disease at the Princeton Medical Center at 12:50 P.M. on Saturday, 14 January 1978. He is survived by Adele Gödel, his wife.

The Bibliography contains a list of several sources for those interested in learning more about the work of Gödel. One of the best of those (Nagel and Newman 1958) states:

> Gödel's findings thus undermined deeply rooted preconceptions and demolished ancient hopes that were being freshly nourished by research on the foundations of mathematics. But his paper was not altogether negative. It introduced into the study of foundation questions a new technique of analysis comparable in its nature and fertility with the algebraic method that René Descartes introduced into geometry. This technique suggested and initiated new problems for logical and mathematical investigation. It provoked a reappraisal, still under way, of widely held philosophies of mathematics, and of philosophies of knowledge in general. [p. 6–7]

*Available as a Dover reprint.

BIBLIOGRAPHY

*Cohen, Paul J. *Set Theory and the Continuum Hypothesis.* New York: W. A. Benjamin, 1966.

Cohen, Paul J., and Reuben Hersh. "Non-Cantorian Set Theory." *Scientific American,* December 1967, pp. 104–16.

*Crossley, J. N., C. J. Nash, C. J. Brickhill, J. C. Stillwell, and N. H. Williams. *What Is Mathematical Logic?* London: Oxford University Press, 1972.

Flint, Peter B. "Kurt Gödel, 71, Dies." *New York Times,* 15 January 1978.

Gödel, Kurt. *On Formally Undecidable Propositions of Principia Mathematica and Related Systems, I.* Reprinted in English in *From Frege to Gödel: A Source Book in Mathematical Logic, 1879–1931,* edited by Jean Van Heijenoort. Cambridge, Mass.: Harvard University Press, 1967.

———. "What Is Cantor's Continuum Problem?" *American Mathematical Monthly* 54 (October 1947): 515–25.

Heijenoort, J. Van. "Gödel's Theorem." In *Encyclopedia of Philosophy.* New York: Crowell Collier & Macmillan, 1967.

Kleene, Stephen C. "The Work of Kurt Gödel." *Journal of Symbolic Logic* 41 (1976): 761–78.

"Milestones." *Time,* 30 January 1978.

Nagel, Ernest, and James R. Newman. *Gödel's Proof.* New York: New York University Press, 1958.

———. "Gödel's Proof." *Scientific American,* June 1956, pp. 71–86.

"Prisoners of Logic and Law—a Certain Genius." *New York Times,* 22 January 1978.

Rosser, Barkley. "An Informal Exposition of Proofs of Gödel's Theorems and Church's Theorem." *Journal of Symbolic Logic* 4 (1939): 56–61.

Sweenburne, Richard. *Space and Time.* New York: St. Martin's Press, 1968.

Whitehead, A. N., and Bertrand Russell. *Principia Mathematica.* Cambridge: At the University Press, 1910, 1912, 1913.

Wiebe, Richard. "Gödel's Theorem (Part I)." *Two Year College Journal of Mathematics* 6 (May 1975): 13–17.

———. "Gödel's Theorem (Part II)." *Two Year College Journal of Mathematics* 6 (September 1975): 4–7.

21

Thinking the Unthinkable:
The Story of Complex Numbers
(with a Moral)

ISRAEL KLEINER

*T*HE USUAL DEFINITION of complex numbers, either as ordered pairs (*a*, *b*) of real numbers or as "numbers" of the form *a* + *bi*, does not give any indication of their long and tortuous evolution, which lasted about three hundred years. I want to describe this evolution very briefly because I think some lessons can be learned from this story, just as from many other such stories concerning the evolution of a concept, result, or theory. These lessons have to do with the impact of the history of mathematics on our understanding of mathematics and on our effectiveness in teaching it. But more about the moral of this story later.

Birth

This story begins in 1545. What came earlier can be summarized by the following quotation from Bhaskara, a twelfth-century Hindu mathematician (Dantzig 1967):

> The square of a positive number, also that of a negative number, is positive; and the square root of a positive number is two-fold, positive and negative; there is no square root of a negative number, for a negative number is not a square.

Reprinted from *Mathematics Teacher* 81 (Oct., 1988): 583–92; with permission of the National Council of Teachers of Mathematics. The author would like to acknowledge financial assistance from the Social Sciences and Humanities Research Council of Canada.

Jerome Cardan (1501–1576)

From *A Portfolio of Eminent Mathematicians*, ed. David Eugene Smith (Chicago: Open Court, 1896)

In 1545 Jerome Cardan, an Italian mathematician, physician, gambler, and philosopher, published a book entitled *Ars Magna* (The great art), in which he described an algebraic method for solving cubic and quartic equations. This book was a great event in mathematics. It was the first major achievement in algebra since the time, 3000 years earlier, when the Babylonians showed how to solve quadratic equations. Cardan, too, dealt with quadratics in his book. One of the problems he proposed is the following (Struik 1969):

> If some one says to you, divide 10 into two parts, one of which multiplied into the other shall produce . . . 40, it is evident that this case or question is impossible. Nevertheless, we shall solve it in this fashion.

Cardan then applied his algorithm (essentially the method of completing the square) to $x + y = 10$ and $xy = 40$ to get the two numbers $5 + \sqrt{-15}$ and $5 - \sqrt{-15}$. Moreover, "putting aside the mental tortures involved" (Burton 1985), Cardan formally multiplied $5 + \sqrt{-15}$ by $5 - \sqrt{-15}$ and obtained 40. He did not pursue the matter but concluded that the result was "as subtle as it is useless" (NCTM 1969). Although eventually rejected, this event was nevertheless historic, since it was the first time ever that the square root of a negative number was explicitly written down. And, as Dantzig (1985) has observed, "the mere writing down of the impossible gave it a symbolic existence."

In the solution of the cubic equation, square roots of negative numbers had to be reckoned with. Cardan's solution for the cubic $x^3 = ax + b$ was given as

$$x = \sqrt[3]{\frac{b}{2} + \sqrt{\left(\frac{b}{2}\right)^2 - \left(\frac{a}{3}\right)^3}}$$

$$+ \sqrt[3]{\left(\frac{b}{2}\right) - \sqrt{\left(\frac{b}{2}\right)^2 - \left(\frac{a}{3}\right)^3}},$$

the so-called Cardan formula. When applied to the historic example $x^3 = 15x + 4$, the formula yields

$$x = \sqrt[3]{2 + \sqrt{-121}} + \sqrt[3]{2 - \sqrt{-121}}.$$

Although Cardan claimed that his general formula for the solution of the cubic was inapplicable in this case (because of the appearance of $\sqrt{-121}$), square roots of negative numbers could no longer be so lightly dismissed. Whereas for the quadratic (e.g., $x^2 + 1 = 0$) one could say that no solution exists, for the cubic $x^3 = 15x + 4$ a real solution, namely $x = 4$, does exist; in fact, the two other solutions, $-2 \pm \sqrt{3}$, are also real. It now remained to reconcile the formal and "meaningless" solution

$$x = \sqrt[3]{2 + \sqrt{-121}} + \sqrt[3]{2 - \sqrt{-121}}$$

of $x^3 = 15x + 4$, found by using Cardan's formula, with the solution $x = 4$, found by inspection. The task was undertaken by the hydraulic engineer Rafael Bombelli about thirty years after the publication of Cardan's work.

Bombelli had the "wild thought" that since the radicands $2 + \sqrt{-121}$ and $2 - \sqrt{-121}$ differ only in sign, the same might be true of their cube roots. Thus, he let

$$\sqrt[3]{2 + \sqrt{-121}} = a + \sqrt{-b}$$

and

$$\sqrt[3]{2 - \sqrt{-121}} = a - \sqrt{-b}$$

and proceeded to solve for a and b by manipulating these expressions according to the established rules for real variables. He deduced that $a = 2$ and $b = 1$ and thereby showed that, indeed,

$$\sqrt[3]{2 + \sqrt{-121}} + \sqrt[3]{2 - \sqrt{-121}}$$

$$= (2 + \sqrt{-1}) + (2 - \sqrt{-1}) = 4$$

(Burton 1985). Bombelli had thus given meaning to the "meaningless." This event signaled the birth of complex numbers. In his own words (*Leapfrogs* 1980):

> It was a wild thought, in the judgement of many; and I too was for a long time of the same opinion. The whole matter seemed to rest on sophistry rather than on truth. Yet I sought so long, until I actually proved this to be the case.

Of course, breakthroughs are achieved in this way—by thinking the unthinkable and daring to present it in public.

The equation $x^3 = 15x + 4$ represents the so-called irreducible case of the cubic, in which all three solutions are real yet they are expressed (by Cardan's formula) by means of complex numbers. To resolve the apparent paradox of cubic equations exemplified by this type of equation, Bombelli developed a calculus of operations with complex numbers. His rules, in our symbolism, are $(\pm 1)i =$

$\pm i$, $(+i)(+i) = -1$, $(-i)(+i) = +1$, $(\pm 1)(-i) = \mp i$, $(+i)(-i) = +1$, and $(-i)(-i) = -1$. He also considered examples involving addition and multiplication of complex numbers, such as $8i + (-5i) = +3i$ and

$$(\sqrt[3]{4 + \sqrt{2i}})(\sqrt[3]{3 + \sqrt{8i}}) = \sqrt[3]{8 + 11\sqrt{2i}}.$$

Bombelli thus laid the foundation stone of the theory of complex numbers.

Many textbooks, even at the university level, suggest that complex numbers arose in connection with the solution of quadratic equations, especially the equation $x^2 + 1 = 0$. As indicated previously, the cubic rather than the quadratic equation forced the introduction of complex numbers.

Growth

Bombelli's work was only the beginning of the saga of complex numbers. Although his book *L'Algebra* was widely read, complex numbers were shrouded in mystery, little understood, and often entirely ignored. Witness Simon Stevin's remark in 1585 about them (Crossley 1980):

> There is enough legitimate matter, even infinitely much, to exercise oneself without occupying oneself and wasting time on uncertainties.

Similar doubts concerning the meaning and legitimacy of complex numbers persisted for two and a half centuries. Nevertheless, during that same period complex numbers were extensively used and a considerable amount of theoretical work was done. We illustrate this work with a number of examples.

As early as 1620, Albert Girard suggested that an equation of degree n may have n roots. Such statements of the fundamental theorem of algebra were, however, vague and unclear. For example, René Descartes, who coined the unfortunate word "imaginary" for the new numbers, stated that although one can imagine that every equation has as many roots as is indicated by its degree, no (real)

numbers correspond to some of these imagined roots.

The following quotation, from a letter in 1673 from Christian Huygens to Gottfried von Leibniz in response to the latter's letter that contained the identity

$$\sqrt{1 + \sqrt{-3}} + \sqrt{1 - \sqrt{-3}} = \sqrt{6},$$

was typical of the period (Crossley 1980):

> The remark which you make concerning . . . imaginary quantities which, however, when added together yield a real quantity, is surprising and entirely novel. One would never have believed that
>
> $$\sqrt{1 + \sqrt{-3}} + \sqrt{1 - \sqrt{-3}}$$
>
> make $\sqrt{6}$ and there is something hidden therein which is incomprehensible to me.

Gottfried Wilhelm von Leibniz (1646–1716)

From *A Portfolio of Eminent Mathematicians*, ed. David Eugene Smith (Chicago: Open Court, 1896)

Leibniz, who spent considerable time and effort on the question of the meaning of complex numbers and the possibility of deriving reliable results by applying the ordinary laws of algebra to them, thought of them as "a fine and wonderful refuge of the divine spirit—almost an amphibian between being and non-being" (*Leapfrogs* 1980).

Complex numbers were widely used in the eighteenth century. Leibniz and John Bernoulli used imaginary numbers as an aid to integration. For example,

$$\int \frac{1}{x^2 + a^2}\, dx = \int \frac{1}{(x+ai)(x-ai)} dx$$

$$= -\frac{1}{2ai} \int \left(\frac{1}{x+ai} - \frac{1}{x-ai} \right) dx$$

$$= -\frac{1}{2ai}$$

$[\log(x + ai) - \log(x - ai)]$. This use, in turn, raised questions about the meaning of the logarithm of complex as well as negative numbers. A heated controversy ensued between Leibniz and Bernoulli. Leibniz claimed, for example, that $\log i = 0$, arguing that $\log(-1)^2 = \log 1^2$, and hence $2\log(-1) = 2\log 1 = 0$; thus $\log(-1) = 0$, and hence $0 = \log(-1) = \log i^2 = 2\log i$, from which it follows that $\log i = 0$. Bernoulli opted for $\log i = (\pi i)/2$; this equation follows from Euler's identity $e^{\pi i} = -1$, which implies that $\log(-1) = \pi i$ and hence that $\log i = \frac{1}{2}\log(-1) = (\pi i)/2$, although this argument is not the one that Bernoulli used. The controversy was subsequently resolved by Leonhard Euler (*Leapfrogs* 1978).

Complex numbers were used by Johann Lambert for map projection, by Jean D'Alembert in hydrodynamics, and by Euler, D'Alembert, and Joseph-Louis Lagrange in incorrect proofs of the fundamental theorem of algebra. (Euler, by the way, was the first to designate $\sqrt{-1}$ by i.)

Euler, who made fundamental use of complex numbers in linking the exponential and trigonometric functions by the formula $e^{ix} = \cos x + i \sin x$, expressed himself about them in the following way (Kline 1972):

> Because all conceivable numbers are either greater than zero, less than zero or equal to zero, then it is clear that the square root of negative numbers cannot be included among the possible numbers.... And this circumstance leads us to the concept of such numbers, which by their nature are impossible and ordinarily are called imaginary or fancied numbers, because they exist only in the imagination.

Even the great Carl Friedrich Gauss, who in his doctoral thesis of 1797 gave the first essentially

Leonhard Euler (1707–1783)

From *A Portfolio of Eminent Mathematicians*, ed. David Eugene Smith (Chicago: Open Court, 1908)

correct proof of the fundamental theorem of algebra, claimed as late as 1825 that "the true metaphysics of $\sqrt{-1}$ is elusive" (Kline 1972).

It should be pointed out that the desire for a logically satisfactory explanation of complex numbers became manifest in the latter part of the eighteenth century, on philosophical, if not on utilitarian, grounds. With the advent of the Age of Reason in the eighteenth century, when mathematics was held up as a model to be followed, not only in the natural sciences but in philosophy as well as political and social thought, the inadequacy of a rational explanation of complex numbers was disturbing.

The problem of the logical justification of the laws of operation with negative and complex numbers also beame a pressing pedagogical issue at, among other places, Cambridge University at the turn of the nineteenth century. Since mathematics was viewed by the educational institutions as a paradigm of rational thought, the glaring inadequacies in the logical justification of the operations with negative and complex numbers became untenable. Such questions as, "Why does $2 \times i + i = 2$?" and "Is $\sqrt{ab} = \sqrt{a}\sqrt{b}$ true for negative a and b?" received no satisfactory answers. In fact, Euler, in his text of the 1760s on algebra, claimed $\sqrt{-1}\sqrt{-4} = \sqrt{4} = +2$ as a possible result. Robert Woodhouse opined in 1802 that since imaginary numbers lead to right conclusions, they must have

Karl Friedrich Gauss (1777–1855)

From *A Portfolio of Eminent Mathematicians*, ed. David Eugene Smith (Chicago: Open Court, 1908)

a logic. Around 1830 George Peacock and others at Cambridge set for themselves the task of determining that logic by codifying the laws of operation with numbers. Although their endeavor did not satisfactorily resolve the problem of the complex numbers, it was perhaps the earliest instance of "axiomatics" in algebra.

By 1831 Gauss had overcome his scruples concerning complex numbers and, in connection with a work on number theory, published his results on the geometric representation of complex numbers as points in the plane. Similar representations by the Norwegian surveyor Caspar Wessel in 1797 and by the Swiss clerk Jean-Robert Argand in 1806 went largely unnoticed. The geometric representation, given Gauss's stamp of approval, dispelled much of the mystery surrounding complex numbers. In the next two decades further development took place. In 1833 William Rowan Hamilton gave an essentially rigorous algebraic definition of complex numbers as pairs of real numbers. (To Hamilton the complex number (a, b) consisted of a pair of "moments of time," since he had earlier defined real numbers, under Immanuel

Kant's influence, as "moments of time.") In 1847 Augustin-Louis Cauchy gave a completely rigorous and abstract definition of complex numbers in terms of congruence classes of real polynomials modulo $x^2 + 1$. In this, Cauchy modeled himself on Gauss's definition of congruences for integers (Kline 1972).

Maturity

By the latter part of the nineteenth century all vestiges of mystery and distrust of complex numbers could be said to have disappeared, although a lack of confidence in them persisted among some textbook writers well into the twentieth century. These authors would often supplement proofs using imaginary numbers with proofs that did not involve them. Complex numbers could now be viewed in the following ways:

1. Points or vectors in the plane
2. Ordered pairs of real numbers
3. Operators (i.e., rotations of vectors in the plane)
4. Numbers of the form $a + bi$, with a and b real numbers
5. Polynomials with real coefficients modulo $x^2 + 1$
6. Matrices of the form
$$\begin{bmatrix} a & b \\ -b & a \end{bmatrix},$$
with a and b real numbers
7. An algebraically closed, complete field (This is an early twentieth-century view.)

Although the preceding various ways of viewing the complex numbers might seem confusing rather than enlightening, it is of course commonplace in mathematics to gain a better understanding of a given concept, result, or theory by viewing it in as many contexts and from as many points of view as possible.

The foregoing descriptions of complex numbers are not the end of the story. Various developments in mathematics in the nineteenth century enable us to gain a deeper insight into the role of

William Rowan Hamilton (1805–1865)

complex numbers in mathematics and in other areas. Thus, complex numbers offer just the right setting for dealing with many problems in mathematics in such diverse areas as algebra, analysis, geometry, and number theory. They have a symmetry and completeness that is often lacking in such mathematical systems as the integers and real numbers. Some of the masters who made fundamental contributions to these areas say it best: The following three quotations are by Gauss in 1801, Riemann in 1851, and Hadamard in the 1890s, respectively:

> Analysis . . . would lose immensely in beauty and balance and would be forced to add very hampering restrictions to truths which would hold generally otherwise, if . . . imaginary quantities were to be neglected. (Birkhoff 1973)

> The origin and immediate purpose of the introduction of complex magnitudes into mathematics lie in the theory of simple laws of dependence between variable magnitudes expressed by means of operations on magnitudes. If we enlarge the scope of applications of these laws by assigning to the variables they involve complex values, then there appears an otherwise hidden harmony and regularity. (Ebbinghaus 1983)

> The shortest path between two truths in the real domain passes through the complex domain. (Kline 1972)

The descriptions of such developments are rather technical. Only the barest of illustrations can be given:

(1) In algebra, the solution of polynomial equations motivated the introduction of complex numbers: Every equation with complex coefficients has a complex root—the so-called fundamental theorem of algebra. Beyond their use in the solution of algebraic polynomial equations, the complex numbers offer an example of an algebraically closed field, relative to which many problems in linear algebra and other areas of abstract algebra have their "natural" solution.

(2) In analysis, the nineteenth century saw the development of a powerful and beautiful branch of mathematics, namely complex function theory. We have already seen how the use of complex numbers gave us deeper insight into the logarithmic, exponential, and trigonometric functions. Moreover, we can evaluate real integrals by means of complex function theory. One indication of the efficacy of the theory is that a function in the complex domain is infinitely differentiable if once differentiable. Such a result is, of course, false in the case of functions of a real variable (e.g., $f(x) = x^{4/3}$).

(3) The complex numbers lend symmetry and generality in the formulation and description of various branches of geometry, for example, Euclidean, inversive, and non-Euclidean. Thus, by the introduction of ideal points into the plane any two circles can now be said to intersect at two points. This idea aids in the formulation and proof of many results. As another example, Gauss used the complex numbers to show that the regular polygon of seventeen sides is constructible with straightedge and compass.

(4) In number theory, certain Diophantine equations can be solved neatly and conceptually by the use of complex numbers. For example, the equation $x^2 + 2 = y^3$, when expressed as $(x + \sqrt{2}i)(x - \sqrt{2}i) = y^3$, can readily be solved, in integers, using properties of the complex domain consisting of the set of elements of the form $a + b\sqrt{2}i$, with a and b integers.

(5) An elementary illustration of Hadamard's dictum that "the shortest path between two truths in the real domain passes through the complex domain" is supplied by the following proof that the product of sums of two squares of integers is again a sum of two squares of integers; that is,

$$(a^2 + b^2)(c^2 + d^2) = u^2 + v^2,$$

for some integers u and v. For,

$$(a^2 + b^2)(c^2 + d^2)$$
$$= (a + bi)(a - bi)(c + di)(c - di)$$
$$= [(a + bi)(c + di)][(a - bi)(c - di)]$$
$$= (u + vi)(u - vi)$$
$$= u^2 + v^2.$$

Try to prove this result without the use of complex numbers and without being given the u and v in terms of a, b, c, and d.

In addition to their fundamental uses in mathematics, some of which were previously indicated, complex numbers have become a fixture in science and technology. For example, they are used in quantum mechanics and in electric circuitry. The "impossible" has become not only possible but indispensable.

The Moral

Why the history of mathematics? Why bother with such "stories" as this one? Edwards (1974) puts it in a nutshell:

> Although the study of the history of mathematics has an intrinsic appeal of its own, its chief raison d'être is surely the illumination of mathematics itself.

My colleague Abe Shenitzer expresses it as follows:

> One can *invent* mathematics without knowing much of its history. One can *use* mathematics without knowing much, if any, of its history. But one cannot

have a mature appreciation of mathematics without a substantial knowledge of its history.

Such appreciation is essential for the teacher to possess. It can provide him or her with insight, motivation, and perspective—crucial ingredients in the making of a good teacher. Of course, whether this story has succeeded in achieving these objectives in relation to the complex numbers is for the reader to judge. However, beyond the immediate objective of lending insight, this story and others like it may furnish us with a slightly better understanding of the nature and evolution of the mathematical enterprise. It addresses such themes or issues as the following:

(1) *The meaning of number in mathematics.* Complex numbers do not fit readily into students' notions of what a number is. And, of course, the meaning of number has changed over the centuries. This story presents a somewhat better perspective on this issue. It also leads to the question of whether numbers beyond the complex numbers exist.

(2) *The relative roles of physical needs and intellectual curiosity as motivating factors in the development of mathematics.* In this connection it should be pointed out that the problem of the solution of the cubic, which motivated the introduction of complex numbers, was *not* a practical problem. Mathematicians already knew how to find approximate roots of cubic equations. The issue was to find a theoretical algebraic formula for the solution of the cubic—a question without any practical consequences. Yet how useful did the complex numbers turn out to be! This is a recurring theme in the evolution of mathematics.

(3) *The relative roles of intuition and logic in the evolution of mathematics.* Rigor, formalism, and the logical development of a concept or result usually come at the end of a process of mathematical evolution. For complex numbers, too, first came *use* (theoretical rather than practical), then *intuitive understanding*, and finally *abstract justification.*

(4) *The nature of proof in mathematics.* This question is related to the preceding item. But although (3) addresses the evolution of complex numbers in its broad features, this item deals with local questions of proof and rigor in establishing various results about complex numbers (cf., e.g., the derivation of the value of logi by von Leibniz and Bernoulli). One thing is certain—what was acceptable as a proof in the seventeenth and eighteenth centuries was no longer acceptable in the nineteenth and twentieth centuries. The concept of proof in mathematics has evolved over time, as it is still evolving, and not necessarily from the less to the more rigorous proof (cf. the recent proof, by means of the computer, of the four-color conjecture). Philip Davis goes a step further in outlining the evolution of mathematical ideas (Davis 1965):

> It is paradoxical that while mathematics has the reputation of being the one subject that brooks no contradictions, in reality it has a long history of successful living with contradictions. This is best seen in the extensions of the notion of number that have been made over a period of 2500 years. From limited sets of integers, to infinite sets of integers, to fractions, negative numbers, irrational numbers, complex numbers, transfinite numbers, each extension, in its way, overcame a contradictory set of demands.

(5) *The relative roles of the individual and the environment in the creation of mathematics.* What was the role of Bombelli as an individual in the creation of complex numbers? Cardan surely had the opportunity to take the great and courageous step of "thinking the unthinkable." Was the time perhaps not ripe for Cardan, but ripe for Bombelli about thirty years later? Is it the case, as John Bolyai stated, that "mathematical discoveries, like springtime violets in the woods, have their season which no human can hasten or retard" (Kline 1972)? This conclusion certainly seems to be borne out by many instances of independent and simultaneous discoveries in mathematics, such as the geometric representation of complex numbers by

George Pólya (1887–1985)
Courtesy of Birkhäuser Boston

Wessell, Argand, and Gauss. The complex numbers are an interesting case study of such questions, to which, of course, we have no definitive answers.

(6) *The genetic principle in mathematics education.* What are the sources of a given concept or theorem? Where did it come from? Why would anyone have bothered with it? These are fascinating questions, and the teacher should at least be aware of the answers to such questions. When and how he or she uses them in the classroom is another matter. On this matter George Pólya (1962) says the following:

> Having understood how the human race has acquired the knowledge of certain facts or concepts, we are in a better position to judge how the human child should acquire such knowledge.

Can we not at least have a better appreciation of students' difficulties with the concept of complex numbers, having witnessed mathematicians of the first rank make mistakes, "prove" erroneous theorems, and often come to the right conclusions for insufficient or invalid reasons?

Some Suggestions for the Teacher

Let me conclude with some comments on, and suggestions for, the use of the history of mathematics in the teaching of mathematics, in particular with reference to complex numbers. Many of the points are implicit in the preceding story.

(1) I first want to reiterate what I view as the major contribution of this story for the teacher. Pólya (1962) puts it very well:

> To teach effectively a teacher must develop a feeling for his subject; he cannot make his students sense its vitality if he does not sense it himself. He cannot share his enthusiasm when he has no enthusiasm to share. How he makes his point may be as important as the point he makes; he must personally feel it to be important.

The objective of my story, then, is to give the teacher some feeling for complex numbers, to imbue him or her with some enthusiasm for complex numbers.

When it comes to suggestions for classroom use, it cannot be overemphasized that these are only suggestions. The teacher, of course, can better judge when and how, at what level, and in what context to introduce and relate historical material to the discussion at hand. The introduction of historical material can, however, convey to the student the following important lessons, which are usually not imparted through the standard curriculum.

(2) Mathematics is far from a static, lifeless discipline. It is dynamic, constantly evolving, full of failures as well as achievements.

(3) Observation, analogy, induction, and intuition are the initial and often the more natural ways of acquiring mathematical knowledge. Rigor and proof usually come at the end of the process.

(4) Mathematicians usually create their subject without thought of practical applications. The latter, if any, come later, sometimes centuries later. This point relates to "immediate relevance" and to "instant gratification," which students often seek from any given topic presented in class.

(5) We must, of course, supply the student with "internal relevance" when introducing a given concept or result. This point brings us to the important and difficult issue of motivation. To some students the applications of a theorem are appealing; to others, the appeal is in the inner logical structure of the theorem. A third factor, useful but often neglected, is the source of the theorem: How did it arise? What motivated mathematicians to introduce it? With complex numbers, their origin in the solution of the cubic, rather than the quadratic, should be stressed. Cardan's attempted division of ten into two parts whose product is forty reinforces this point. How much further one continues with the historical account is a decision better made by the teacher in the classroom, bearing in mind the lessons that should be conveyed through this or similar historical material.

(6) Historical projects deriving from this story about complex numbers can be given to able students as topics for research and presentation to, say, a mathematics club. Possible topics are the following:

(a) The logarithms of negative and complex numbers.

(b) What is a number? That is, discuss the evolution of various number systems and the evolution of our conception of what a number is.

(c) Hypercomplex numbers (e.g., the quaternions). Their discovery is another fascinating story.

(d) Gauss's congruences of integers and Cauchy's congruences of polynomials. The latter lead to a new definition (description) of complex numbers.

(e) An axiomatic characterization of complex numbers (see (7) under the heading "Maturity"). In this connection we ought to discuss the notion of characterizing a mathematical system, and thus the concept of isomorphism. (Cf. the various equivalent descriptions of complex numbers discussed previously.)

(7) Many elementary and interesting illustrations of Hadamard's comment demonstrate that "the shortest path between two truths in the real

domain passes through the complex domain." We are referring to elementary results from various branches of mathematics, results whose statements do not contain complex numbers but whose "best" proofs often use complex numbers. One such example was given previously. Some others by Cell (1950), Jones (1954), and the NCTM (1969) can be found in the Bibliography.

BIBLIOGRAPHY

Birkhoff, Garrett. *A Source Book in Classical Analysis.* Cambridge, Mass.: Harvard University Press, 1973.

Burton, David M. *The History of Mathematics.* Boston: Allyn & Bacon, 1985.

Cell, John W. "Imaginary Numbers." *Mathematics Teacher* 43 (December 1950):394–96.

Crossley, John N. *The Emergence of Number.* Victoria, Australia: Upside Down A Book Co., 1980.

Dantzig, Tobias, *Number—the Language of Science.* New York: The Free Press, 1930, 1967.

Davis, Philip J. *The Mathematics of Matrices.* Waltham, Mass.: Blaisdell Publishing Co., 1965.

Ebbinghaus, Heinz-Dieter, et al. *Zahlen.* Heidelberg: Springer-Verlag, 1983.

Edwards, Charles H. *The Historical Development of the Calculus.* New York: Springer-Verlag, 1974.

Flegg, Graham. *Numbers—Their History and Meaning.* London: Andre Deutsch, 1983.

Jones, Phillip S. "Complex Numbers: An Example of Recurring Themes in the Development of Mathematics—I–III." *Mathematics Teacher* 47 (February, April, May 1954): 106–14, 257–63, 340–45.

Kline, Morris. *Mathematical Thought from Ancient to Modern Times.* New York: Oxford University Press, 1972.

Leapfrogs: Imaginary Logarithms. Fordham, Ely, Cambs., England: E. G. Mann & Son, 1978.

Leapfrogs: Complex Numbers. Fordham, Ely, Cambs., England: E. G. Mann & Son, 1980.

McClenon, R. B. "A Contribution of Leibniz to the History of Complex Numbers." *American Mathematical Monthly* 30 (November 1923): 369–74.

Nagel, Ernest. "Impossible Numbers: A Chapter in the History of Modern Logic." *Studies in the History of Ideas* 3 (1935): 429–74.

National Council of Teachers of Mathematics. *Historical Topics for the Mathematics Classroom.* Thirty-first Yearbook. Washington, D.C.: The Council, 1969.

Pólya, George. *Mathematical Discovery.* New York: John Wiley & Sons, 1962.

*Sondheimer, Ernest, and Alan Rogerson. *Numbers and Infinity—an Historical Account of Mathematical Concepts.* New York: Cambridge University Press, 1981.

Struik, Dirk J. *A Source Book in Mathematics, 1200–1800.* Cambridge, Mass.: Harvard University Press, 1969.

Windred, G. "History of the Theory of Imaginary and Complex Quantities." *Mathematical Gazette* 14 (1930?): 533–41.

*Available as a Dover reprint.

22

The Development of Modern Statistics

DALE E. VARBERG

First Lecture

THAT AREA OF STUDY which we now call statistics has only recently come of age. While its origins may be traced back to the eighteenth century, or perhaps earlier, the first really significant developments in the theory of statistics did not occur until the late nineteenth and early twentieth centuries, and it is only during the last thirty years or so that it has reached a full measure of respectability. It was antedated by the theory of probability and has its roots embedded in this subject. In fact, any serious study of statistics must of necessity be preceded by a study of probability theory, for it is in the latter subject that the theory of statistics finds its foundation and fountainhead.

The word *statistik* was first used by Gottfried Achenwall (1719–1772), a lecturer at the University of Göttingen.[1] He is sometimes referred to as the "Father of Statistics"—perhaps mistakenly, since he was mainly concerned with the description of interesting facts about his country.

Our English word "statistic" means different things to different people. To the man on the street, statistics is the mass of figures that the expert on any subject uses to support his contentions—it's "what you use to prove anything by." To the more sophisticated person, the word may evoke some notion of the procedures which are used to condense and interpret a collection of data, such as the computing of means and standard deviations. But to the practitioner of the craft, statistics is the art of making inferences from a body of data, or, more generally, the science of making decisions in the face of uncertainty.

Statisticians concern themselves with answering such questions as: Is this particular lot of manufactured items defective? Is there a connection between smoking and cancer? Will Kennedy win the next election? In answering these questions, it is necessary to reason from the specific to the general, from the sample to the population. Therefore, any conclusions reached by the statistician are not to be accepted as absolute certainties. It is, in fact, one of the jobs of the statistician to give some measure of the certainty of the conclusions he has drawn.

It should not be inferred from this lack of certainty that the mathematics of statistics is nonrigorous. The mathematics that forms the basis of statistics stems from probability theory and has a firm axiomatic foundation and rigorously proved theorems.

If we conceive of statistics as the science of drawing inferences and making decisions, it is

Reprinted from *Mathematics Teacher* 57 (Apr., 1963): 252–57; 57 (May, 1963): 344–48; with permission of the National Council of Teachers of Mathematics. This is the text of two lectures on the history of statistics given by Professor Varberg at a National Science Foundation Summer Institute for High School Mathematics Teachers held at Bowdoin College during the summer of 1962. Notes for these lectures were taken by Alvin K. Funderburg.

appropriate to date its beginnings with the work of Sir Francis Galton (1822–1911) and Karl Pearson (1857–1936) in the late nineteenth century. Starting here, modern statistical theory has developed in four great waves of ideas, in four periods, each of which was introduced by a pioneering work of a great statistician.[2]

The first period was inaugurated by the publication of Galton's *Natural Inheritance* in 1889. If for no other reason, this book is justly famous because it sparked the interest of Karl Pearson in statistics. Until this time, Pearson had been an obscure mathematician teaching at University College in London. Now the idea that all knowledge is based on statistical foundations captivated his mind. Moving to Gresham College in 1890 with the chance to lecture on any subject that he wished, Pearson chose the topic: "the scope and concepts of modern science." In his lectures he placed increasingly stronger emphasis on the statistical foundation of scientific laws and soon was devoting most of his energy to promoting the study of statistical theory. Before long, his laboratory became a center in which men from all over the world studied and went back home to light statistical fires. Largely through his enthusiasm, the scientific world was moved from a state of disinterest in statistical studies to a situation where large numbers of people were eagerly at work developing new theory and gathering and studying data from all fields of knowledge. The conviction grew that the analysis of statistical data could provide answers to a host of important questions.

An anecdote, related by Helen Walker,[3] of Pearson's childhood illustrates in a vivid way the characteristics which marked his adult career. Pearson was asked what was the first thing he could remember. "Well," he said, "I do not know how old I was, but I was sitting in a high chair and I was sucking my thumb. Someone told me to stop sucking it and said that unless I do so the thumb would wither away. I put my two thumbs together and looked at them a long time. 'They look alike to me,' I said to myself, 'I can't see that the thumb I suck is any smaller than the other. I wonder if she could be lying to me.'"

We have here in this simple story, as Helen Walker points out, "rejection of constituted authority, appeal to empirical evidence, faith in his own interpretation of the meaning of observed data, and finally imputation of moral obliquity to a person whose judgment differed from his own." These were to be prominent characteristics throughout Pearson's whole life.

This first period, then, was marked by a change in attitude toward statistics, a recognition of its importance by the scientific world. But, in addition to this, many advances were made in statistical technique. Among the technical tools invented and studied by Galton, Pearson, and their followers were the standard deviation, correlation coefficient, and the chi square test.

About 1915, a new name appeared on the statistical horizon, R. A. Fisher (1890–). His paper of that year on the exact distribution of the sample correlation coefficients ushered in the second period of statistical history and was followed by a whole series of papers and books which gave a new impetus to statistical inquiry. One author has gone so far as to credit Fisher with half of the statistical theory that we use today. Among the significant contributions of Fisher and his associates were the development of methods appropriate for small samples, the discovery of the exact distributions of many sample statistics, the formulation of logical principles for testing hypotheses, the invention of the technique known as analysis of variance, and the introduction of criteria for choice among various possible estimators for a population parameter.

The third period began about 1928 with the publication of certain joint papers by Jerzy Neyman and Egon Pearson, the latter a son of Karl Pearson. These papers introduced and emphasized such concepts as "Type II" error, power of a test, and confidence intervals. It was during this period that industry began to make widespread application of statistical techniques, especially in connection with quality control. There was increasing interest in taking of surveys with consequent attention to the theory and technique of taking samples.

We date the beginning of the fourth period with the first paper of Abraham Wald (1902–1950) on the now often used statistical procedure—sequential sampling. This paper of 1939 initiated a deluge of papers by Wald, ended only by his untimely death in an airplane crash when at the height of his powers. Perhaps Wald's most significant contribution was his introduction of a new way of looking at statistical problems, what is known as statistical decision theory. From this point of view, statistics is regarded as the art of playing a game, with nature as the opponent. This is a very general theory, and, while it does lead to formidable mathematical complications, it is fair to say that a large share of present-day research statisticians have found it advantageous to adopt this new approach.

Having given this brief bird's-eye view of statistical history, we move to a discussion of some of the most basic of statistical concepts. For this purpose it will be convenient to refer to a table showing the heights and weights of twelve people

FIGURE I

Table of Heights and Weights

Individual	X	Y
1	60	110
2	60	135
3	60	120
4	62	120
5	62	140
6	62	130
7	62	135
8	64	150
9	64	145
10	70	170
11	70	185
12	70	160

FIGURE 2

Frequency Diagram

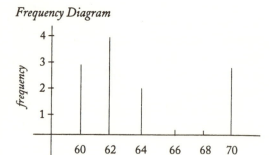

(Fig. 1). The height X is shown in inches; the weight Y is shown in pounds.

To get some feeling for such a collection of data, it is clearly desirable to display the data pictorially. William Playfair (1759–1823) of England is usually given credit for introducing the idea of graphical representation into statistics. His writings, mostly on economics, were illustrated with extremely good graphs, histograms, bar diagrams, etc. In our problem, the data is most simply represented by means of what is called a frequency diagram.

We have shown such a diagram for the height X (Fig. 2). A similar diagram for Y would be easy to construct. While such pictures do help our intuition, we need more than this if we are to treat the data mathematically. We need mathematical measures which describe the data precisely.

Among the most important of such measures are the measures of central tendency. The earliest of these, actually dating back to the Greeks, is the arithmetic mean μ, which for a discrete variable X, such as we have in our example, is defined by

$$\mu_X = (1/n) \sum_{i=1}^{n} x_i.$$

Here x_i denotes a value of the variable X, and n is the size of the population. In our example, the mean μ_X of the heights is 63.83; the mean μ_Y of the weights is 141.67.

To understand the significance of this concept, we rewrite the definition in the form

$$\mu_X = (1/n) \sum x_j f_j.$$

Here f_j stands for the frequency of occurrence of the value x_j and the summation extends over the distinct values of the variable X. Consider now a weightless rod on which there is a scale running through the range of the variable X, and suppose that at x_j is attached a mass of size f_j/n. This gives a system of total mass 1, which will have μ_X as its center of mass, that is, the system will balance on a fulcrum placed at μ_X. In the case of the heights, the system would look as in Figure 3. This interpretation of the mean will be helpful later when we consider the notion of a continuously distributed variable.

While the concept is probably quite old, it was not until 1883 that the median was introduced into statistics by Galton as a second measure of central tendency.[4] The median is simply the middle value of the distribution in the case of an odd number of values and is the average of the two middle values otherwise. The median height in our example is 62.

Another measure of central tendency is the mode, introduced by Karl Pearson around 1894. The mode is the most frequently occurring value, if there is one. In the case where two or more values occur with equal frequency, the mode is not well defined. In the example, the mode of heights is again 62.

If the distribution of a variable X is exactly symmetrical, i.e., if its frequency diagram is exactly symmetrical about a vertical line, then the mean, median, and mode (if there is a mode) will agree. The reader should be able to convince himself that the converse is false by constructing a non-symmetrical distribution for which the mean, median, and mode agree.

FIGURE 3

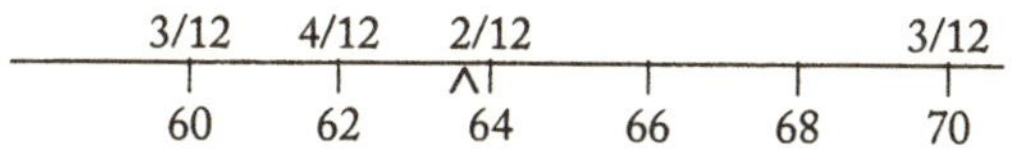

For most purposes, certainly for theoretical purposes, the mean is the most useful measure of central tendency, although admittedly it may take much calculation to get it. The median does, however, have a property which is sometimes advantageous. It is not as subject to distortion due to a few extreme values. For example, if in the table of heights of twelve persons, one of the 70-inch persons were exchanged for a 90-inch person, the mean would be changed considerably while the median would be unaffected.

We next consider measures of dispersion, i.e., measures of how the data spreads out about the mean. Perhaps the first such measure was the probable error introduced by Bessel in 1815 in connection with problems in astronomy. Most commonly used today is the standard deviation σ, this terminology due to Karl Pearson (1894). It is defined for a discrete variable X by

$$\sigma_X = \left[(1/n \sum_{i=1}^{n} (x_i - \mu_X)^2 \right]^{1/2}.$$

Inspection of this formula reveals that σ tends to be large when the data is widely dispersed, small when the data clusters about the mean.

To introduce the next notion, which is correlation, we refer back to the table of heights and weights (Fig. 1). Inspection of the data reveals that these two variables are somehow related. Even common sense tells us that tall people should generally weigh more than short people. Graphically, this relationship can be portrayed by means of what is called a scatter diagram, this being merely a plot of the data in the Cartesian plane (see Fig. 4). The relationship, if linear, will be indicated by a tendency of the points to simulate a straight line.

In the late nineteenth century, Sir Francis Galton asked whether such a relationship between two sets of data could be measured, and he introduced the notion of correlation. It was Karl Pearson, however, who gave us our present coefficient of correlation ρ defined by

FIGURE 4

Scatter Diagram

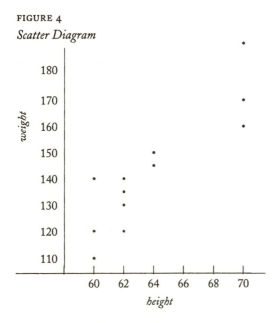

$$\rho = (1/n\sigma_X\sigma_Y \sum_{i=1}^{n} (x_i - \mu_X)\,(y_i - \mu_Y).$$

It is a matter of simple algebra to show that ρ ranges between −1 and +1. A value of zero indicates no linear relationship; +1 indicates that the data lies on a straight line of positive slope; −1 means that the data lies on a line of negative slope; values near ±1 suggest a strong linear relationship, while values near zero are characteristic of little such relationship. In our example, r is about 0.9. It should be emphasized that r is a measure of linear relationship. The data may lie on a circle, in which case $|\rho|$ will be very small. However, in this case the variables would certainly be related, albeit not linearly related.

Sir Francis Galton, prominent in our discussion thus far, was a cousin of Charles Darwin and did some statistical work for him. His interest in correlation has already been mentioned—it was his writings on this subject which turned the brilliant Karl Pearson toward the study of statistics. But Galton will be remembered most vividly by teachers because he was the first to suggest the use of the normal curve in connection with problems of grading.

The normal curve, which actually dates back at least to Abraham De Moivre in 1733, is a highly useful concept to statistics. It is determined by the equation

$$f(x) = (1/\sqrt{2\pi}\ \sigma)\ \exp\ [\ -(x-\mu)^2/2\sigma^2].$$

Here μ and σ are parameters which turn out to be the mean and standard deviation. The normal curve is often thought of roughly as any "bell shaped" curve. However, this is inaccurate, for other functions, such as $g(x) = [\pi\,(1 + x^2)]^{-1}$, also have graphs which are bell shaped and yet lack completely the qualities which make the normal curve so useful. While the definition of the normal curve given above may appear complicated, from the point of view of the mathematician it is one of the simplest and best behaved of all curves. Figure 5 pictures a special normal curve.

If the area under the normal curve from $-\infty$ to $+\infty$ were to be calculated by integration, it would be found to be 1. Approximately two-thirds of this area lies between points one standard deviation to the left and one standard deviation to the right of the mean. The probability that a normal variable assumes values on any interval $a \leq x \leq b$ is equal to the area above this interval and under the corresponding normal curve. Areas under the normal curve for various intervals are tabulated in any standard book of mathematical tables.

Earlier, in the discussion of discrete distribution, it was shown that the mean could be

FIGURE 5

The Normal Curve (μ = 1, σ = 1)

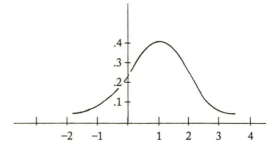

interpreted as the center of mass of a system of discrete masses of total mass 1. The normal distribution described above is an example of a continuous distribution. Reasoning by analogy, we may associate with the normal distribution an idealized continuous rod of mass 1 running indefinitely far in both directions with density varying according to the function f which determines the normal curve. From calculus, the center of mass μ of such a rod would be given by

$$\mu = \int_{-\infty}^{\infty} xf(x)dx.$$

This is, in fact, the formula which we use to define the mean of a continuous distribution. Perhaps surprisingly, not every continuous distribution has a mean, for the above integral may fail to converge. This is the case, for example, for the Cauchy distribution, determined by the equation $g(x) = [\pi(1+x^2)]^{-1}$, as the reader may verify.

Recalling the formula in the discrete case, it is natural to define the standard deviation for a continuous distribution by

$$\sigma = \left[\int_{-\infty}^{\infty} (x - \mu)^2 f(x)dx \right]^{1/2}.$$

It is a matter of a fairly simple integration to check that if these formulas are used to calculate the mean and standard deviation for the normal distribution, they turn out to be the two parameters μ and σ respectively.

Second Lecture

In attempting to answer questions concerning large populations, it becomes necessary, from a practical standpoint, to work with *samples* from the population. The population parameters, such as the mean μ and the standard deviation σ are then unknown. Now assuming that a sample is properly selected (how to do this is an important statistical problem),

it should be possible to get good estimates of the population parameters from the sample. R. A. Fisher has introduced criteria for judging whether a sample statistic is a good estimator for a population parameter. We shall not discuss these criteria. Suffice it to say that if $(x_1, x_2, \ldots, x_n)$ denotes a sample from a population with mean μ and standard deviation σ, then the sample mean $\bar{x}$ and sample standard deviation σ, defined by

$$\bar{x} = (1/n) \sum_{i=1}^{n} x_i$$

and

$$s = \left[(1/n) \sum_{i=1}^{n} (x_i - \bar{x})^2 \right]^{1/2}$$

respectively, turn out to satisfy most of the criteria that Fisher has given.[5]

Now if we were to take many samples from a population and compute $\bar{x}$ for each of them, we would get many different values, but presumably these values would tend to cluster about the population mean μ. Looked at this way, $\bar{x}$ is a variable which is distributed in some fashion. This raises an important question. Given a certain distribution of the population variable, how is $\bar{x}$, the sample mean, distributed? A theorem, which we state without proof, partially answers this question.

THEOREM. *If the population variable is normally distributed with mean μ and standard deviation σ, then $\bar{x}$ is normally distributed with the same mean μ but with standard deviation $\overline{\sigma} / \sqrt{n}$, n being the size of the sample.*

FIGURE 6

Normal distributions of X and x for n = 10, 20

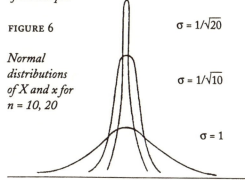

$\sigma = 1/\sqrt{20}$

$\sigma = 1/\sqrt{10}$

$\sigma = 1$

Recalling the significance of the standard deviation, we conclude, not unexpectedly, that as the sample size is made larger and larger the values of $\bar{x}$ tend to cluster more and more closely about μ. This is illustrated pictorially in Figure 6.

A grave limitation arises in the use of this theorem since it is doubtful whether any population variables in real life are exactly normally distributed and many are not even approximately so. We are saved by the most famous theorem in probability theory and one of the most famous in all of mathematics. It is known as the central limit theorem. One form of it is the following:

THEOREM. *If the population variable is distributed in any fashion whatever so long as it has a mean μ and a standard deviation σ, then $\bar{x}$ is approximately normally distributed with mean μ and standard deviation $\sigma/\sqrt{n}$. The word approximately is used in the sense that the distribution of $\bar{x}$ approaches nearer and nearer to the normal distribution as n becomes larger and larger.*

The central limit theorem has a long history. It was proved in its first form about 1733 by Abraham De Moivre for the case of tossed pennies where there are only two possible outcomes. The form which we have stated is due to J. W. Lindeberg in 1922.[6] In recent years, Russian mathematicians have generalized this theorem to its absolute extreme, giving necessary and sufficient conditions for $\bar{x}$ to have the normal distribution as its limiting distribution.

To show one of the uses that the statistician makes of the central limit theorem, consider the following typical problem taken from a well known textbook by Hoel.[7] "A manufacturer of string has found from past experience that samples of a certain type of string have a mean breaking strength of 15.6 pounds with a standard deviation of 2.2 pounds. A time-saving change in the manufacturing process of this string is tried. A sample of 50 pieces is then taken, for which the mean breaking strength turns out to be only 14.5 pounds. On the basis of this sample, can it be concluded that the new process has had a harmful effect on the strength of the string?"

A statistician views the problem this way. It is necessary to test the hypothesis H_0 that $\mu = 15.6$ against the alternative hypothesis H_1 that $\mu < 15.6$. It will be assumed that X, the breaking strength, still has a standard deviation of 2.2, although the possibility certainly exists that the change in manufacturing process has changed the standard deviation. We may now appeal to the central limit theorem, for, no matter how X is distributed, we know that $\bar{x}$ is approximately normally distributed with mean μ and standard deviation $\sigma/\sqrt{n}$, or equivalently that $z = (\bar{x} - \mu)\sqrt{n}/\sigma$ is approximately normally distributed with mean 0 and standard deviation 1. Now if a standard normal table is consulted, we find that a value of 14.5 for $\bar{x}$ is so far away from 15.6 that under hypothesis H_0 the probability of a value this small occurring is only .0002. It seems safe therefore to reject H_0 and accept H_1. The tables also reveal that, if we follow the usual statistical procedure of rejecting H_0 whenever a value occurs which is so small that under H_0 it would occur only 5 per cent of the time, we should reject H_0 for a value of $\bar{x}$ less than 15.09. Any value less than 15.09 falls in the so-called critical region (see Fig. 7).

We mention again the possibility of error due to the assumption that the standard deviation σ has not changed under the new process. Actually, σ is no longer a known parameter. We can, however, calculate s, the sample standard deviation. In 1908, a chemist, William Gosset, writing under the pseudonym of "Student" discovered the distribution of the variable $t = (\bar{x} - \mu)\sqrt{n-1}/s$ (note σ is replaced by s). He showed that if X is normally distributed, the t has the "Student's t" distribution with $n - 1$ degrees of freedom. This distribution is of sufficient importance so that it is tabulated in

FIGURE 7

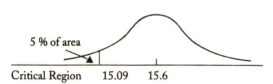

5 % of area

Critical Region   15.09   15.6

most statistical tables. The requirement that X be normally distributed is a stringent one. However, even if X is only approximately normally distributed, statisticians have found it advantageous to use the t distribution when σ is unknown, especially if n is small. If n is large, the difference between s and σ may be expected to be small; hence there is less need to consult the t distribution tables.

We have credited Gosset with the discovery of the t distribution in 1908. However, it should be pointed out that Gosset's results were really guesses and were not rigorously verified until about 1926 by R. A. Fisher, a bright young statistician who had since 1915 begun to play an increasingly influential role.

Fisher is said to have been a precocious child,[8] having mastered such subjects as spherical trigonometry at an early age. He was attracted to the physical sciences and received his B.A. in astronomy from Cambridge in 1912. The theory of errors of astronomy eventually led to his interest in statistical problems. We have picked the date 1915 for his introduction to the statistical world because in that year he published a paper on the distribution of the sample correlation coefficient. This paper initiated a study of the exact distribution of various sample statistics, at which Fisher has been eminently successful. In this study he has been guided by his tremendous geometrical intuition, often arriving at results which have later been proved correct only by the concentrated effort of some of the world's top-flight mathematicians.

Fisher has made many other contributions. Previously, we mentioned his introduction of criteria for judging whether a sample statistic is a good estimator for a population parameter. These include the concepts of consistency, efficiency, and sufficiency, which took form in an impressive memoir of 1921. Also falling in this category is his introduction of maximum likelihood estimators.

In 1919, Fisher left his job of teaching mathematics in public schools to work at the Rothamsted Agricultural Experiment Station. Here he developed sampling techniques and randomization

procedures which are now used all over the world. Two of his books, *Statistical Methods for Research Workers*, published in 1925, and *Design of Experiments*, published in 1935, have had a tremendous influence. The second chapter of this latter book is included in the *World of Mathematics*.[9] In this very charming article, Fisher tells of a woman who claims that she can tell whether or not the milk in her tea was poured before or after the tea was poured, and goes on to describe an experiment designed to prove or disprove her claim.

The theory of testing and estimation received a fresh start with the work of Jerzy Neyman and E. S. Pearson that began about 1928. This resulted in a reformulation and modification of many of the ideas originally introduced by Fisher. To illustrate one of their contributions, we refer back to the problem of the string manufacturer. Our conclusion was that, if a sample led to an $\bar{x}$ which was less than 15.09, the hypothesis H_0 was to be rejected (see Fig. 7). Neyman and Pearson raised questions of the following variety. Why do we take as critical region the region to the left of 15.09? Why do we not choose a "two tailed" critical region with two and one-half per cent of the area at the extreme left and two and one-half per cent of the area at the the extreme right of the distribution? What criteria should be used in choosing critical regions? Must intuition be used or can sound mathematics be brought into the picture? There resulted the formulation of a table which revealed that two distinct types of error were possible, and which Neyman and Pearson named Type I and Type II errors (see Fig. 8).

FIGURE 8

Types of error in testing hypotheses

	H_0 True	H_1 True
$\bar{x} \geq 15.09$ Accept H_0	Correct decision	Type II error
$\bar{x} < 15.09$ Accept H_1	Type I error	Correct decision

Neyman and Pearson were able to sum up their findings in this guiding principle: *Among all tests (critical regions) possessing the same Type I error, choose one for which the size of the Type II error is as small as possible.* While the application of this principle is reasonably complicated, the influence of Neyman and Pearson has made this principle and the related notion of power function important statistical concepts, and a general mathematical theory dealing with this kind of problem has been developed.

No modern account of the history of statistics would be complete without mention of the name of Abraham Wald. Beginning about 1939 and continuing until his death in 1950 in an airplane crash when at the height of his powers, Wald profoundly influenced the direction of statistical theory. His followers are among the leaders of the field today.

Wald was born in Romania of orthodox Jewish ancestry.[10] Because of his religion, he was denied certain educational privileges and found it necessary to study on his own. A clue to his mathematical ability is contained in the fact that as a result of his own studies he was able to make valuable suggestions concerning Hilbert's *Foundations of Geometry*, suggestions which were incorporated in the seventh edition of that book.

Wald later attended the University of Vienna and received his Doctor's Degree after taking only three courses. At this period of time in the political history of Austria, he was denied the privilege of academic work and had to accept a private position helping a banker broaden his knowledge of higher mathematics. He became interested in the theory of economics. Later he became a close associate of the economist, Morgenstern, who collaborated with John von Neumann in the field of game theory.

Wald came to the United States during the dark days before World War II, during which his parents and sisters were eventually to lose their lives in a gas chamber. His interest in economics led him to study statistics, and he soon became an outstanding theoretical statistician. Perhaps his most important con-

tribution was the introduction of a new way of looking at statistics—statistical decision theory. From this point of view, statistics is regarded as the art of playing a game with nature as the opponent. Much of the recent theoretical study has been in this direction and even elementary textbooks are now beginning to adopt this point of view.

Abraham Wald made many other important contributions, but space allows us to cite only one—sequential analysis. Although perhaps not original with him, the theory was certainly developed by him. The technique was judged to be so important in the way of minimizing sampling procedures in manufacturing processes that it was highly classified by the military during World War II.

We shall illustrate the idea of sequential analysis in connection with the problem of quality control in industry. Before the introduction of sequential methods, the standard procedure was to take a fixed size sample from each lot of manufactured items and then accept or reject the lot on the basis of the number of defectives in the sample. This procedure ignores the fact that information about the lot can be obtained from the rate at which defective items occur in the sampling process.

In sequential sampling, three possible states which may occur during the sampling process are recognized: (1) early appearance of defective items

FIGURE 9

Sequential sampling

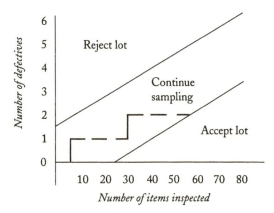

Number of items inspected

leading to a quick decision to reject the lot, or (2) early appearance of a lack of defective items leading to a quick decision to accept the lot, or (3) inconclusive evidence in which case the sampling process should be continued. This is illustrated pictorially in Figure 9 for a particular example. The guidelines separating the three regions are determined according to the Type I and Type II errors which are allowed. In this example, a decision to *accept* would be made after examining the sixtieth item.

From the diagram, it is clear that while it may be possible to make a quick decision to accept or reject, it is also possible to stay in the intermediate region for a long time or perhaps even indefinitely. Wald, however, showed that, with probability 1, a decision to accept or reject will be made in a finite number of steps. Actual experience has shown that sequential sampling usually results in a saving of about fifty per cent in the cost of sampling when compared with the traditional fixed-sample-size procedures.

NOTES

Editor's note: Notes 5–10 were originally numbered 1–6. Likewise, Figures 6–9 were originally numbered 1-4.

1. Walker, Helen M., *Studies in the History of Statistical Method* (Baltimore: The William and Wilkins Company, 1929). The book has been of great value in the preparation of this lecture, especially in connection with statistical history before 1900.

2. In dividing the history of statistics into four periods, we are following Helen M. Walker, "The Contributions of Karl Pearson," *Journal of the American Statistical Association*, LIII (1958), 11–22.

3. *Ibid.*, p. 13.

4. Actually, Gustav Fechner had employed this measure under the name *der Centralwerth* in 1874 and had given a description of its properties. Galton's use of the concept appears to go back as early as 1869, but the name *median* is first used by him in 1883.

5. In order to be unbiased, n should be replaced by $n-1$ in the formula for s.

6. J. W. Lindeberg, "Eine neue Herleitung des Exponentialgesetzen in der Wahrscheinlich-keitsrechnung," *Mathematisch Zeitschrift*, XV (1922), 211–25.

7. Paul G. Hoel, *Introduction to Mathematical Statistics*, 2nd Edition (New York: John Wiley and Sons, Inc., 1954), p. 103.

8. For a discussion of the life and contributions of Fisher, see Harold Hotelling, "The Impact of R. A. Fisher on Statistics," *Amer. Stat. Assoc.*, XLVI (1951), 35–46.

* 9. James R. Newman, ed., *The World of Mathematics*, Vol. 3 (New York: Simon and Schuster, 1956), pp. 1512–21.

10. Karl Menger, "The Formative Years of Abraham Wald and his Work in Geometry," *Ann. Math. Stat.*, 23:1 (1952), 14–20.

*Available as a Dover reprint.

HISTORICAL EXHIBIT 4

A Chronological Outline of the Evolution of Computing Devices

SINCE VERY EARLY TIMES, humans have sought to simplify the tasks of numerical record keeping and computation by using physical devices. These devices have varied greatly in scope and context from the adaption of personal body parts such as fingers and toes for simple 1:1 tallying to the use of inanimate objects, notched sticks, and knotted cords for numerical record keeping to the operation of complex mechanical and electrical machines for scientific calculations. Today's high-speed digital computers can trace their origins to the finger manipulations of our ancient ancestors. The path of this evolution is outlined in the list of accomplishments and names of individuals given below.

Date	Accomplishment or Event
?	Use of hands and fingers to communicate numerical facts
ca. 30,000 B.C.	Tally bones recovered from European sites
ca. 8000 B.C.	Clay tokens used in Babylonia for numerical record keeping
ca. 600 B.C.	Abacus used in Classical Greece
ca. 500 B.C.	Computing rods and counting board introduced in China
ca. A.D. 600	With the collapse of Imperial Rome, use of the column abacus dominates European computing. In European setting evolves into a line abacus computing table which remains in use until time of late Renaissance.
ca. A.D. 1400	Quipu used by Inca people of pre-Columbian America
A.D. 1614	John Napier develops logarithms, invents Napier's rods for carrying out multiplication
1620	Edmund Gunter develops logarithmic scale basis for slide rule capable of performing four basic operations
1623	Wilhelm Schickard invents computing machine that can perform four operations
1642	Blaise Pascal builds gear-driven computer that can perform addition and subtraction with six-digit numbers
1671	Gottfried Wilhelm Leibniz refines design of gear computer to include "stepped cylinders" allowing for operation of multiplication and division by repeated additions or subtraction.
1673	Sir Samuel Morland invents multiplying machines in England
1805	Joseph Marie Jacquard develops punch card input for textile looms
1820	Thomas de Colmar standardizes design for mechanical computing machines

Date	Accomplishment or Event
1830	Charles Babbage conceives of great computing engines capable of 26-digit computations. Babbage's designs incorporate specifications of modern digital computers, i.e., input, processing unit, output.
1875	Frank Baldwin obtains American patent for popular calculating machine
1941	Konrad Zuse developes Z3, a relay calculator possessing 64 word memory
1944	Automatic Sequence Controlled Calculator (ASCC) built at Harvard
1945	Electronic Numerical Integrator and Computer (ENIAC) begins operation at University of Pennsylvania, contains 18,000 vacuum tubes, performs 360 multiplication/sec.
	John von Neumann develops Electronic Discrete Variable Calculator (EDVAC) at University of Pennsylvania.
1947	Transistor developed at Bell Laboratories
1951	U.S. Census Bureau accepts delivery of Remington Rand UNIVAC 1. The computer contains 5000 vacuum tubes and performs 1000 calculations/sec.
1953	Magnetic core memory introduced into computers
1957	Fortran programming language introduced
1959	Concept of integrated circuits conceived by Robert Noyce
1960	Cobol language introduced
1964	IBM 360 marketed, employs binary addressing, introduces cheap feasible time-sharing and virtual memory.
	Basic language introduced
1968	First Ph.D. in computer science awarded at University of Pennsylvania
1969	UNIX operating system introduced
	Edgar Codd proposes relational database model to IBM
	Intel develops microprocessor
1970	Floppy disc introduced
1971	Pascal language introduced
	First pocket calculators appear
1975	Microcomputers marketed
1976	Cray 1 supercomputer becomes operational
	Kenneth Appel and Wolfgang Haken resolve 4-color conjecture using a computer
1980	Ada language introduced
1985	The Connection Machine developed by Thinking Machines Corporation, a highly parallel supercomputer possessing 65,536 processors
1988	Computer networking well established

Date	Accomplishment or Event
1990	Introduction of Windows 3.0 by Bill Gates and Microsoft
1993	Intel Pentium released
1995	Java Script development announced by Netscape
1996	Netscape Navigator 2.0 released
1997	IBM's Deep Blue beats Chess Champion Gary Kasparov
1999	Linux Kernel 2.2.0 released important operating system in Unix world
2001	Apple releases Mac OSX
2002	Edgar Dykstra dies—noted for shortest path algorithm (1956)
2003	Sir Tim Berners-Lee knighted in recognition of creation of World Wide Web

23

ENIAC: The First Computer

AARON STRAUSS

THE ELECTRONIC DIGITAL COMPUTER is a war baby. The high-speed calculating and data-processing machines have led to the development of new ideas and methods in pure mathematics; they have made possible technological progress leading to interplanetary space travel; and they administer activities of daily life from traffic control and department store sales to industrial payrolls and income tax collection. The computer is truly a major contribution of American mathematics, science, and technology. Yet the stimulus for the production of ENIAC—as so often has been true throughout history—was the demand for ever more sophisticated military methods.

The First World War introduced the first long-range weapons of twentieth-century warfare: cannons that could fire at targets miles away and planes that could drop bombs on targets from miles above. But the gunners and bombardiers who used these weapons needed detailed information for aiming the artillery or bombs. Gunners needed to know angles of elevation for various combinations of target distance and crosswind conditions. Bombardiers needed directions on release time as a function of altitude and air speed. The new weapons received only limited use in World War I. But in the early stages of World War II,

long-distance artillery and bombers played a major role in the fighting.

In 1942, the United States Army sent a young mathematician, Herman H. Goldstine, to the Moore School of Engineering at the University of Pennsylvania to supervise the preparation of firing and bombing tables. Computations required for accurate tables were enormous, and the Moore School was chosen for the project because it had been the home of a great deal of research and development on computing machines (mostly analog computers) during the 1930s. To calculate the trajectory of a typical projectile fired from a large cannon could take over 750 multiplications and many other calculations. These computations would require twelve man-hours using desk calculators, an hour or two using the fastest digital computers available, or at least twenty minutes using the fastest analog computers. Thus hours, or even days, were required to furnish a firing table for just one type of gun. Goldstine and two engineers, John Mauchly and J. P. Eckert, proposed to the Army that the Moore School develop a digital computer that would be at least 100 times faster than the fastest computer then available. The Army officially accepted the proposal in June 1943. By 1946, a computer called ENIAC (Electronic Numerical Integrator and Computer) had been constructed with the capability of performing the trajectory calculations mentioned above in less than ten seconds. More important, the design was soon improved so that subsequent models were a thousand times faster than ENIAC.

Reprinted from *Mathematics Teacher* 69 (Jan., 1976): 66–72; with permission of the National Council of Teachers of Mathematics.

Author's note: I thank Jack Goldhaber, Jack Minker, Chris Rapalus, and Werner Rheinboldt for their helpful suggestions concerning this article.

Historical Evolution of the Computer Concept and Technology

Though ENIAC was produced in the United States through efforts directed by Goldstine, Mauchly, and Eckert, the project built on important ideas of many mathematicians and engineers—from both earlier and contemporary times. The calculating capabilities of ENIAC and its successors were undoubtedly dreamed of by mathematicians for centuries. But the main outline of the computer story is given by the following chronology.

1623:   Calculator of Wilhelm Schickard (Germany).

1642:   Calculator of Blaise Pascal (France).

1673:   Calculator of Gottfried Leibniz (Germany).

1805:   Automated loom of Joseph Marie Jacquard (France).

1822–62:   Calculators of Charles Babbage (England).

1853:   Calculator of Pehr Georg Scheutz (Sweden).

1874:   Calculator of Martin Wiberg (Sweden).

1890:   Use of punch card tabulators in U.S. Census.

1929+:   IBM Computing Bureau at Columbia University.

1937+:   IBM Computer project at Harvard University.

1937+:   Computer project at Bell Telephone Laboratories.

1937+:   Electronic computer ideas of John V. Atanasoff (Iowa).

1943:   Contract to build ENIAC (Philadelphia).

1944:   John von Neumann joins ENIAC project.

1945:   Planning of EDVAC and von Neumann's "First Draft."

1946:   ENIAC finished and operating.

1947:   Conversion of ENIAC into a stored program computer.

1946–58:   Institute for Advanced Study computer project (Princeton).

1951:   Remington Rand UNIVAC.

1953:   IBM 701.

The first calculators of Schickard, Pascal, and Leibniz were hand operated—that is, mechanical. They were about the size of a television set. In principle, they operated like the little plastic adding machines that some people use when shopping in supermarkets, except that some also did multiplication and division. These calculators came before the Industrial Revolution, and they showed that at least this one human activity of calculating could be mechanized. But the calculators were slow and certainly not widely used.

Many fundamental principles of modern computers were suggested in the work of an Englishman, Charles Babbage (1792–1871). He proposed using steam power (the "modern" energy source in those days) to run the calculator and using punched cards for entering data into the machine (an idea apparently acquired from Jacquard, who automated weaving by using punched cards to guide weaving patterns in the looms). Babbage and his wife, Lady Lovelace (daughter of Lord Byron), described programs for their machines in ways that influenced programs developed 100 years later. Unfortunately, nineteenth-century technology could not match the ideas Babbage produced, and he never completely finished building a calculator.

Most of Babbage's ideas were developed for what he called an "Analytical Engine." A much simpler calculator described by Babbage, called a "Difference Engine," was actually built by Scheutz and later improved by Wiberg. In fact, Wiberg used his machine to calculate tables of logarithms, which he then entered in the Philadelphia Exhibition of *1876*!

The first connection between the United States and calculating machines seems to have been with the use of tabulators in the 1890 U.S. Census. Herman Hollerith developed a tabulating system in his doctoral work at Columbia University during the 1880s, improved that system, and eventually used it to tabulate the entire 1890 census within one month after the returns were in, an unprecedented feat. Hollerith's tabulators read punch cards and were electromechanical—that is, they used electric power to throw relays and switches. They were about the size of a dining-room hutch cabinet, or a bedroom chest of drawers. Not only did these machines mark the entrance of the United States into the calculating-machine field, but they were the first to use many of Babbage's advanced ideas and the first to be used for an important, large-scale commercial venture. The use of Hollerith machines spread. For example, in the 1920s, L. J. Comrie (England) used such a machine to calculate tables of the positions of astronomical objects.

One of the early manufacturers of Hollerith machines, the Tabulating Machine Company, later became IBM. Between the two world wars, IBM helped set up calculating laboratories at two American universities. The laboratory at Harvard operated under H. H. Aiken and produced the Mark I, a giant, electromechanical calculator of complex design, similar to another developed at Bell Telephone Laboratory. Both were outmoded by the ENIAC, but they served to introduce many people at IBM, Harvard, and Bell Labs into the computing field. The IBM-Harvard Mark I was about ten feet high, three feet deep, and over fifty feet long. These electromechanical calculators operated in principle like an electric desk calculator or a supermarket cash register, except that they read punch cards, stored numbers, took square roots, and were more complex.

Why weren't these giant electromechanical computers good enough? They were too slow; they could not furnish the data needed by the war effort. The problem was that it took too long for an electric current to throw a relay. Vacuum tubes, on the other hand, could be charged or discharged thousands of times faster than a switch could be thrown. Vacuum tubes are electronic; switches are electromechanical. The modern, quiet, department-store cash registers that click instead of ring are examples of electronic adding and filing machines. The new pocket calculators made by Texas Instruments, Hewlett-Packard, and others are examples of electronic (solid state) calculating machines. However, thousands of tubes were needed for an electronic computer, and much design work was necessary to make the computer reliable enough to

be useful. This was a "hardware" problem. There were also "software" problems of how to communicate with the machine. Pioneering work on hardware problems was done by Atanasoff, whose ideas influenced Mauchly just before Mauchly went to the Moore School in 1941. This brings us to the Moore School/U.S. Army contract to build the ENIAC, which, incidentally, was about the size of the IBM-Harvard Mark I.

About a year after the ENIAC project commenced, the Moore School was visited by one of the most brilliant and versatile mathematicians of this century, John von Neumann. Von Neumann was born and educated in Europe and had emigrated to America in the 1930s to accept a professorship at Princeton and then at Princeton's Institute for Advanced Study. He became interested in the ENIAC project, joined it in the summer of 1944, and made fundamental contributions to both the hardware and software areas. By the fall of 1944, most of the work of designing the ENIAC was finished. This left the designers relatively free to design a second computer while the ENIAC was being constructed. They had many improvements in mind; for example, they wanted to reduce the number (18,000) of vacuum tubes required by the ENIAC. This new machine was called EDVAC (Electronic Discrete Variable Computer), and von Neumann wrote his ideas in a paper called "First Draft of a Report on the EDVAC." Although this paper was not intended for publication, it was widely circulated and was credited by a number of others for showing them how to design a computer.

It is hard to explain, and harder to overstate, the magnitude of von Neumann's contribution. Perhaps his most fundamental contribution was that he described exactly what a computer is, or ought to be. He said it should have a central arithmetic part that is actually a calculating machine, capable of performing the basic arithmetic functions. It should have a control part, which determines how the machine deals with its orders, and in what kind of sequence it obeys its instructions. It should have the ability to read data from the outside world (input) and to inform the outside world of its results (output). And it should have a memory, capable of storing numbers for further use and for storing instructions. In fact this feature of storing its instructions is perhaps what distinguishes a computer from a calculator. Indeed, von Neumann went on to develop the concept of the self-modifying, stored program: some of the instructions tell the computer how to modify some of its other instructions.

After the war the ENIAC/EDVAC group split up. Mauchly and Eckert formed a computer company that eventually produced the UNIVAC, while von Neumann and Goldstine headed a project to continue research work on new computers at the Institute for Advanced Study. Later von Neumann became a consultant to IBM and Goldstine went there full time. The Smithsonian Institution in Washington, D.C., now has an exhibit of these early computers and their calculator predecessors. But the contemporary offspring of ENIAC are in evidence throughout our daily lives.

The Computer and America

In what sense is the computer an *American* contribution to (or from) mathematics? First the ENIAC was the world's first electronic digital computer, and it was designed and built in the United States. Second, von Neumann's "First Draft" was perhaps the most influential single paper written on designing computers. Third, the virtual explosion of computers on the scene in the 1950s was composed of American-made computers.

A more difficult question is, What were the particular features of the United States that caused the first computer to be built here? Most of the earlier developments (before 1890) occurred elsewhere. In the history of computers, two factors stand out: (1) the interplay of industry, government, and academia; and (2) the existence in the 1930s and 1940s of fast-growing electrical and electronic technology and big electrical companies.

Regarding (1): In addition to the IBM-Harvard example mentioned earlier and the lowering of the barriers separating industry, universities, and government during the war, it is interesting to note that the Institute for Advanced Study computer project was financed variously by the Office of Naval Research, Army Ordnance, the Atomic Energy Commission, the Radio Corporation of America, and Princeton University. This type of interplay has come to be the topic of much controversy in recent years.

Regarding (2): The idea of a computer was not totally new. What was new was the sophistication of electronics necessary to achieve the speed and the reliability to make the whole thing worthwhile. Vacuum tubes (electronic) are much faster than relays (electromechanical). For example, the whole field of numerical meteorology was made possible by the speed of modern computers. A model was run on the Institute for Advanced Study computer in 1953 and yielded a twenty-four-hour weather forecast in six minutes. That model took twenty-four hours to run on the ENIAC and, of course, hopelessly longer on its predecessors. The technology was quite highly developed in the United States, as can be exemplified by the electrical, telephone, telegraph, and radio networks, and by big companies such as General Electric, the Radio Corporation of America, and Western Electric.

Some work was done on computers in Germany during the period 1937–45, especially by Konrad Zuse. In fact, Zuse actually built several electromechanical computers and achieved some success in designing a compiler. (A compiler is a language close to a human one, and the computer itself is programmed to translate this language into its own special code, which it then follows.) Why didn't the Germans continue? Here we arrive at another reason why computers developed in the United States: it was the only major industrial nation to escape physical devastation during World War II.

Speculation on the causes of historical events is a hazardous occupation in which one is inevitably limited by incomplete information. In fact, controversy over who should get credit for inventing the computer is manifest in several pending patent suits between early contributors to the field. None-theless, the factors outlined above seem to have had a clear and important role in making the computer an American contribution to mathematics and technology.

The Computer and Mathematics

The development of the computer was sparked by problems in applied mathematics. But the design of effective computers was also facilitated in important ways by theoretical ideas from mathematics, and the availability of modern computing power has stimulated growth of entirely new branches of pure and applied mathematics.

As von Neumann conceived of the computer, it should consist of processor, control, input/output, and memory—operations of which should be guided by a stored program. Mathematics provides the theoretical models of the electronic components that performed these functions. In early computers, the easiest and most reliable way to use electronics to represent numbers was by using, not the *amount* of current, but simply the *direction* of flow—positive or negative, on or off; a switch could be open or closed. Thus it was a natural to use the binary system (base 2) of numeration, rather than the decimal. Most computers after the ENIAC used binary arithmetic, although the ENIAC itself was a decimal machine. Binary arithmetic is an example of Boolean algebra, which, together with other aspects of formal symbolic logic, was the main tool of pure mathematics used in the design of early computers. More recently, concepts from lattice theory, numerical analysis, and combinatorics have become fundamental tools in the design of computers and their software.

The individual aspects of a computer design provide no striking new insight into mathematics, but combined in an electronic system, they provide a calculating and logical tool with incredible speed. The significance of this speed has been stated well by others:

> But why are these great speeds so important? Are they really necessary or have they been devised only to keep the new machine busy?" Let us start the discussion by stating categorically that without the speed made possible only by electronics our modern computerized society would have been impossible: machines that can do as much as ten or twenty or thirty or even a hundred humans are very important but do not revolutionize modern society. They are extremely valuable and help greatly to ease the burden on humans, but they do not make possible an entirely new way of life ... clearly efforts such as are typified by numerical weather prediction would be quite impossible without electronic computers This is then the whole point of the modern machines. It is not simply that they expedite highly tedious, burdensome, and lengthy calculations being done by humans or electromechanical machines. It is that they make possible what could never be done before! The electronic principle did much more than free men from the loss of hours like slaves in the labor of calculating: it also enabled them to conceive and execute what could not ever be done by men alone. It is what made possible putting men on the moon—and bringing others back safely from an abortive trip there. [Goldstine 1972, pp. 145–47]

Another argument that continually arises is that machines can do nothing that we cannot do ourselves, though it is admitted that they can do many things faster and more accurately. The statement is true, but also false. It is like the statement that, regarded solely as a form of transportation, modern automobiles and aeroplanes are no different than walking. One can walk from coast to coast of the U.S. so that statement is true, but is it not also quite false? Many of us fly across the U.S. one or more times each year, once in a while we may drive, but how few of us ever seriously consider walking more than 3000 miles? The reason the statement is false is that it ignores the order of magnitude changes between the three modes of transportation: we can walk at speeds of around 4 miles per hour, automobliles travel typically around 40 miles per hour, while modern jet planes travel at around 400 miles per hour. Thus a jet plane is around two orders of magnitude faster than unaided human transportation, while modern computers are around six orders of magnitude faster than hand computation. It is common knowledge that a change by a single order of magnitude may produce fundamentally new effects in most fields of technology; thus the change by six orders of magnitude in computing have produced many fundamentally new effects that are being simply ignored when the statement is made that computers can only do what we could do for ourselves if we wished to take the time. [Hamming 1964, pp. 1–2]

These order-of-magnitude changes in the speed with which traditional mathematical calculations can be carried out have enhanced the applicability of classical ideas. The computer can quickly calculate tables of sines and cosines, logarithms, square roots, definite integrals, statistical parameters, or roots of equations. When the first computer was completed in 1946, it found immediate use in the Los Alamos thermonuclear project, but also in more peaceful pursuits like studying the distribution of prime numbers.

But the computer's ability to cope with the many calculations required in the solution of complex system probelms has also played a fundamental role in creating important new branches of mathematics. For instance, whereas any reasonable linear programming application is a situation involving dozens of variables subject to equally numerous linear constraints, the calculations required to solve the system are prohibitively complex in the absence of programmable computer power. The application of computers to space travel or weather prediction stimulates further research in the mathematics of these areas. The tremendous number of calculations involved creates "round-off error" problems that have become the basis of a new mathematical specialty called numerical stability. Problems in computer design have led to the study of topics like the theory of algorithms and the mathematics of grammar and language. Areas of mathematics that had been investigated and forgotten have gained new life because of the needs and potential of computers. For instance, numerical analysis, combinatorics, graph theory, network theory, and automata theory are now prominent topics for pure and applied mathematical research. The ability of computers to store and retrieve large amounts of information has prompted extensive investigation of their potential as both medium and manager of instruction in all areas of education.

One of the chief differences between a calculator and a computer is that the latter can be programmed in advance, the program itself can be stored in the computer, and the program can modify itself, just as it can modify numbers. In 1966, almost twenty years after the first stored program computer, it was proved mathematically that self-modifying programs are more powerful than non-self-modifying ones; another, yet undoubtedly not the last, example of mathematics "created" by the computer.

Reflections

The chronicle of developments that produced the first computer in America, with major contributions by Americans, can be viewed simply as a story of impressive mathematical and scientific achievement. But like any thoughtful reading of history, that story raises broader questions about the motivation, process, and impact of progress in our discipline: Why does it seem to take a war to produce scientific advances that could just as well come about without wars? The computer is not the only war baby that has had enormous practical payoff in peaceful pursuits. Radar and rockets were also both products of World War II; and over two thousand years ago Archimedes produced many of his startling mathematical results in response to problems of national defense.

A discussion of the American contribution to any field raises the question, Who is an American? One who is native born? A citizen? A resident? In acknowledging the American contribution to a discipline, where is the dividing line between justifiable pride on the one hand and a kind of America-can-do-everything nationalism on the other? Big Business and the cooperative efforts of government, industry, and universities can produce good (such as the computer) or evil (such as destruction of the environment). How should mathematicians and laymen deal with this problem?

Many further references to specific parts of the computer story appear in Goldstine's fine book, *The Computer from Pascal to von Neumann* (1972), which served as the principal source for the historical development contained in this article. More extensive discussion of software developments may be found in the second-to-last chapter of Goldstine and the first few pages of Rosen (1967).

The texts listed under student references highlight history as well as future implications of computer technology.

BIBLIOGRAPHY

For Teachers

Goldstine, Herman H. *The Computer from Pascal to von Neumann.* Princeton: Princeton University Press, 1972.

Hamming, R. W. "Impact of Computers." *American Mathematical Monthly* 72, pt. 2 "Computers and Computing," (January 1964): 1–7.

Richtmyer, R. D. "The Post-War Computer Development." *American Mathematical Monthly* 72, pt. 2 "Computers and Computing," (January 1964): 8–14.

Rosen, Saul, ed. *Programming Systems and Languages.* New York: McGraw-Hill Book Co., 1967.

von Neumann, John. *The Computer and the Brain.* New Haven: Yale University Press, 1958.

For Students

Eames, Charles, and Ray Eames. *A Computer Perspective.* Cambridge, Mass.: Harvard University Press, 1973.

Feingold, Carl. *Introduction to Data Processing.* Dubuque, Iowa: Wm. C. Brown, 1971.

Kemeny, John G. *Man and the Computer.* New York: Charles Scribner's Sons, 1972.

Martin, James, and Adrian R. D. Norman. *The Computerized Society.* Englewood Cliffs, N.J.: Prentice-Hall, 1970.

Marxner, Ellen. *Elements of Data Processing.* Albany, N.Y.: Litton Educational Publishing, 1971.

Miller, Boulton B. *Computers: A User's Introduction.* Edwardsville, Ill.: Bainbridge, 1974.

Orilia, Lawrence, William M. Fuori, and Anthony D'Arco. *Introduction to Computer Operations.* New York: McGraw-Hill Book Co., 1973.

Sanders, Donald H. *Computers in Society.* New York: McGraw-Hill Book Co., 1973.

Shelly, Gary B., and Thomas J. Cashman. *Introduction to Flowcharting and Computer Programming Logic.* Fullerton, Calif.: Anaheim Publishing Co., 1972.

24

Dynamical Systems:
Birkhoff and Smale

JAMES KAPLAN AND AARON STRAUSS

$\mathcal{B}$EGINNING PERHAPS with Newton in the seventeenth century many brilliant thinkers have attempted to express the observable behavior of the physical world in the form of relatively simple mathematical formulas and equations. Frequently these equations are "differential equations," which relate the position of some object (or objects) to its velocity, acceleration, temperature changes, energy, external forces, and so on. At first, differential equations were used by Newton to describe the motions of both celestial and earthbound objects under the influence of various gravitational forces. His famous equation $F = ma$, the second of Newton's three laws, is really a differential equation. Later, differential equations were used to describe the behavior of certain mechanical systems such as springs, pendulums, and engines; then to describe properties of electrical networks; and recently they were applied to describe some aspects of the twentieth-century fields of atomic physics, solid state electronics, biology, economics, and space travel.

In the late nineteenth century the superb French mathematician Henri Poincaré (1854–1912) initiated a study of differential equations from a more abstract point of view. In this way he hoped to unify the existing theories and shed some light on certain problems in celestial mechanics that had

been puzzling scientists at that time. He studied systems in which the particles moved according to two or three basic mathematical "laws." These laws were general enough to include most of the specific situations already mentioned. These systems came to be known as *dynamical systems*.

Today the study of dynamical systems is an active, exciting field of mathematical research. Perhaps more than any other field, it combines in a basic way the three principal mathematical areas of analysis, algebra, and topology and uses this combination to obtain direct applications to the problems in the physical world mentioned above. Since the time of Poincaré, important theoretical advances in dynamical systems have been made by several American mathematicians. This article focuses on two of them—George David Birkhoff (1884–1944) and Stephen Smale (1930–).

We shall begin with some examples to give the reader more of a feeling for dynamical systems, briefly discuss the accomplishments of Birkhoff and Smale, and finally describe in more detail one major advance made by each.

Examples

Imagine that a sheet of paper is placed over a bar magnet. If a tiny iron ball bearing is placed on the sheet, held at rest for a moment, and then released, the bearing will move under the influence of the magnetic force. If the bearing is first dipped in ink and then released on the paper, it will leave a path

Reprinted from *Mathematics Teacher* 69 (Oct., 1976): 495–501; with permission of the National Council of Teachers of Mathematics.

FIGURE I

Four particles and their integral curves

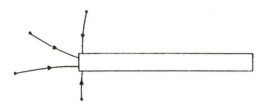

as it moves. If the bearing is again dipped in ink and placed down at exactly the same initial spot, it will have exactly the same motion, that is, it will follow its previous path. If the bearing is redipped and placed at different spots, different paths will be traced (Fig. 1). This is a primitive example of a dynamical system. In the language of dynamical systems, the bearing is a "particle," its movement is the "motion of the particle," each path is a "trajectory," or "integral curve," or "path," and each path is "unique" in the sense that a bearing placed twice in the same spot will follow the same path both times. The collection of all the paths (although we actually draw only a representative sample) is called the "phase portrait" of the dynamical system.

Now suppose that black iron filings are sprinkled on the paper and that the paper is gently tapped to allow the filings to settle into a pattern (Fig. 2). The pattern of black curves will look very much the same as the phase portrait of paths obtained previously. The difference is that no particle traced out the black curves. All the black

FIGURE 2

Black curves representing a vector field

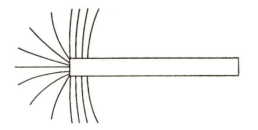

curves were formed simultaneously and directly from the lines of magnetic force. In the language of dynamical systems, the black curves are the "vector field"; they show the "external force" that will produce the motion of particles. The particles will move "along" or "smoothly through" the vector field and trace out paths. Primitive examples similar to this include the motion of a Ping-Pong ball in a mountain stream and the path of a drop of black paint that is dropped into a pail of white paint that in turn is being stirred slowly and uniformly.

Uncontrived examples are more complicated because it is necessary to include velocity as part of the description of a particle's position. (The next two examples require the use of analytic geometry; the reader who has not learned this subject is advised to skip the next two paragraphs.)

Suppose a pendulum swings, and at time t its angle with the vertical is denoted by $x(t)$, or simply x (Fig. 3). The motion is represented by paths in a two-dimensional space. The horizontal axis is x, the vertical axis is the velocity of x, denoted $\dot{x}$. The

FIGURE 3

Simple pendulum

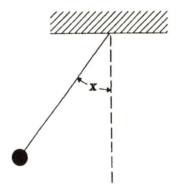

point (2,0) means $x = 2$, $\dot{x} = 0$; point (1, 1) means $x = 1$, $\dot{x} = 1$, and so on. If there is no air resistance, the pendulum has a periodic motion. This is represented by a closed curve around the origin (e.g., a circle) in the x, $\dot{x}$ plane. This is called a periodic

FIGURE 4

Path of a pendulum

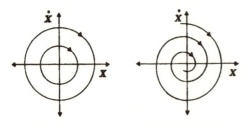

(a) no air resistance   *(b) some air resistance*

path (see Fig. 4a). If there is air resistance, the pendulum swings back and forth, but the amplitude (maximum angle) steadily decreases, and the pendulum approaches a state of rest. This is represented in the x, $\dot{x}$ plane as a spiral contracting toward the origin (see Fig. 4b). The system without air resistance is called a *conservative* dynamical system. This name comes from the fact that in such a system energy is conserved. In the system with air resistance, energy is lost as the pendulum slows down.

In the earth-moon-sun system of three mutually dependent objects, six dimensions are needed, three for the position of each body and three for the velocity of each body. This system seems to be, for all practical purposes, conservative.

In studying any dynamical system the principal goal is to predict the future behavior of the system from information that we have about it at present. This is far from easy; in fact, for a conservative system of three bodies, each of which exerts a force on the other two and in which no simplifying assumptions are made, this goal has not yet been achieved. This problem is called the "three-body problem," and mathematicians have been trying to solve it for hundreds of years. Through attempts to solve this and other problems, many results have been obtained that, although they do not solve the problem, help us to understand dynamical systems. We can now turn to a discussion of the lives of Birkhoff and Smale and to the results they obtained.

Birkhoff and Smale

George David Birkhoff, who was intellectually a disciple of Poincaré, was unquestionably one of the world's premier mathematical intellects. In addition to his work on dynamical systems, which occupied most of his life, he published papers on such diverse problems as the four-color problem, relativity theory, number theory, and even aesthetics. His collected works comprise about a hundred memoirs, nearly three thousand pages, and he also published several books.

One of the first native-born, American-educated American mathematicians, Birkhoff was born in Michigan in 1884 and received his doctorate (classical ordinary differential equations) at the University of Chicago in 1907. He held positions at the University of Wisconsin and Princeton and, for most of his career, Harvard. The high quality of his work was recognized worldwide. In 1918 he received the Querini-Stampalia Prize of the Royal Institute of Science, Arts, and Letters of Venice; in 1923 he received the Bôcher Prize of the American Mathematrical Society; he was awarded the annual prize of the American Association for the Advancement of Science in 1926; he received the Pontifical Academy of Sciences prize in 1933. He was also a member of the National Academy of Sciences, the American Academy of Arts and Sciences, the American Philosophical Society, president of the American Mathematical Society (1925), and president of the American Association for the Advancement of Science (1937).

Stephen Smale received his doctorate (modern differential topology) at the University of Michigan in 1956. He has since held positions at the University of Chicago, the Institute for Advanced Study at Princeton, Columbia University, and is presently at the University of California at Berkeley. Smale's research is on differential topology and global analysis (dynamical systems from a global, or topological, point of view). He has applied his mathematical techniques to problems in electrical circuit theory, economics, and, more recently, biology. In the beginning of his career Smale worked on, among other things, the famous Poincaré conjecture, which proposed that if an object has certain specific properties that spheres have, then that object must in fact be a sphere. This problem in the field of algebraic topology represents an area other than dynamical systems in which both Smale and Poincaré did fine work. Smale solved the Poincaré conjecture in higher dimensions (while sunbathing on a beach, he once said), and for this feat he was awarded the Veblen Prize of the American Mathematical Society in 1964 and the Fields Medal of the International Mathematical Union in 1966. A Nobel Prize is not awarded in mathematics; the Fields Medal is probably its equivalent. It was at this point that Smale began his work on dynamical systems.

Obviously, it would be impossible in a single short article to describe in any detail even a small fraction of the work of two creative and prolific mathematicians of the stature of Birkhoff and Smale. Instead, we shall content ourselves with trying to describe intuitively two mathematical ideas that are important in dynamical systems theory, one of which is associated with each man—the "ergodic theorem" with Birkhoff and "structural stability" with Smale. In a sense, both of these ideas were developed to overcome a common problem in dynamical systems: a lack of completely accurate (i.e., error free) information. In order to understand these ideas we must examine in a little more detail the notion of a dynamical system. For convenience, we shall restrict ourselves to two-dimensional systems. In what follows, the reader who is unfamiliar with analytic geometry should keep in mind the example of the sheet of paper over the bar magnet in Figures 1 and 2.

Dynamical Systems

Consider a particle p that moves in the usual x,y-plane under the influence of certain forces. Suppose that corresponding to each point q in the plane there is associated an arrow $F(q)$ whose tail

FIGURE 5

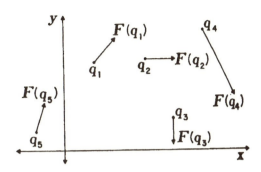

is located at q (see Fig. 5). The direction of the arrow indicates the direction of the external force on a particle located at q, and the length of the arrow denotes the force's magnitude. The collection of all such arrows (or *vectors*) in the plane is called a *vector field*. A particle placed at any point q will be moved by the force $F(q)$, and at each subsequent location it will be acted upon by the force associated with that new location. The path followed by a particle under the influence of the vector field is called an *integral curve*, and the collection of all possible integral curves (which is the same as all possible motions of a particle whose motion is determined by the vector field) constitutes the *phase portrait* of the dynamical system (see Fig. 6).

Recall that our goal in the study of dynamical systems is to be able to predict future behavior from present information. Thus it seems clear that a knowledge of the complete phase portrait of our system is sufficient for our purpose. Given the current location of a particle, we need only follow the integral curve through that point to see where the particle will be at any future instant.

It is at this point that we encounter one of the fundamental problems of science, alluded to earlier: the difficulty of measurement. We can never determine *exactly* the forces (or integral curves) in our vector field, just as we can never determine exactly the current position of a moving particle. We are limited by the accuracy of our measuring devices. Worse, the small errors made when taking

FIGURE 6

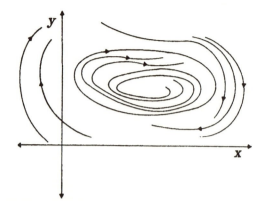

our initial measurements will tend to become magnified as we try to make predictions further and further into the future. In a sense, Birkhoff's ergodic theorem and the notion of structural stability that has come to be associated with Smale are attempts to resolve in different ways the problem of imperfect information.

How would you attempt to resolve the problem if you were faced with it? For instance, suppose you measured the length of a rod on three separate occasions and found its length to be 40.1, 39.9, and 40.0 centimeters. What value would you use for the length of the rod? Probably, you would assume it to be 40.0 centimeters, which is the average of the different measurements.

Now let's return to our dynamical system. Consider again a particle p moving in the x,y-plane. Suppose we wish to know the value of a particular quantity (such as the energy) that depends on the location of p. Such quantities are sometimes called *observables* of the system. We might ask, "What will be the value of the observable at some future time t?" As we have indicated, if we never know the exact position of p, we shall never know the true value of the observable. Instead, by analogy with the previous discussion, we might ask, "As a particle p moves, what will be the average value of our observable?" Such considerations serve as motivation for the subject of *ergodic theory*.

The average value of our observable as the particle moves along an integral curve is an example of a *time average*, since the values of the observable we are averaging are computed at each future time. There is another kind of average, too—a *space average*. We could consider all possible locations of the particle p at one instant of time (say, the present) and then average the value of the observable corresponding to each of these locations.

Perhaps an example can clarify this distinction. Suppose we want to determine the average energy possessed by a molecule of air in a certain room. Here, energy is the observable. There are two ways to do this. One method is to follow a molecule forever, record its energy, and compute its average energy over its lifetime. This is a *time average*. The second method is to compute the energies of every molecule of air in the room and find the average value of all those energies. This is a *space average*. Birkhoff's ergodic theorem states that, under certain technical assumptions, for conservative dynamical systems (systems in which energy is neither gained nor lost) the space average of any observable equals its time average. Notice how this achieves our fundamental goal of predicting future behavior from present information. We can use the present average value of the observable over all of space to predict its average future value. Provided we are willing to settle for average values of conservative systems, we can resolve the problem of imperfect information.

Another attempt to resolve this same problem is the notion of structural stability. To understand this idea, let's first imagine that the phase portrait of our dynamical system is drawn on a flat rubber sheet. Any new phase portrait that is obtainable from the old one simply by stretching or pulling the rubber sheet is called *topologically equivalent* to it (see Fig. 7). Topologically equivalent phase portraits exhibit roughly the same behavior. Notice that we are not permitted to cut a hole in the rubber, nor can we fold it over. Figure 4 contains

FIGURE 7

Topologically equivalent phase portraits

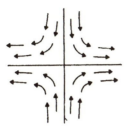

two phase portraits that are *not* topologically equivalent.

With this in mind, let's return for a moment to the picture of our vector field (see Fig. 8). Imagine that we alter each of the arrows by a "little bit." We can change its length or its direction. In so doing we obtain a new vector field that, in turn, will generate a new phase portrait. We say that our original dynamical system is *structurally stable* if every new dynamical system obtained by "slight" modification of the vector field has a phase portrait that is topologically equivalent to the old one. This idea was originally used by Andronov, a Soviet engineer, and Pontryagin, a Soviet mathematician, but it has since become associated with Smale because of his outstanding work on it.

If we are willing to restrict ourselves to structurally stable systems, then imperfect information is not crucial, since our lack of exact knowledge will result in a slightly modified dynamical system that is topologically equivalent to the original one.

FIGURE 8

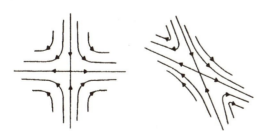

Thus we can use our approximation to predict future behavior, secure in the knowledge that our results will be approximately correct. "Most" dynamical systems in the plane are, in fact, structurally stable. The same is unfortunately not true in higher dimensions.

Reflections

Needless to say, for reasons of space, we are forced to omit from our discussion several other American mathematicians who also made important contributions to dynamical system theory. Furthermore, this field is by no means exclusively American. We have concentrated on the American contributions because of the Bicentennial.

Our chief regret is that we cannot refer the reader to an elementary, yet complete, exposition of dynamical systems. The subject is an advanced one, requiring a lot of knowledge from many areas of mathematics. We can only hope that this article has served to give the reader at least a taste of the flavor of this rich subject.

Finally, we cannot resist a comment about the roles of these two mathematical giants, Birkhoff and Smale, in the development of mathematics in America. Prior to Birkhoff, Americans had to study in Europe to learn significant mathematics. Birkhoff was one of the "new breed," American born and American educated. This American tradition is now firmly established, verified by the biographies of Smale and his contemporaries and even from the fact that many European mathematicians now come to the United States to complete part of their education. But in another sense, a tradition seemed to end with Birkhoff's death—the tradition that mathematicians did fundamental and important work in areas other than pure mathematics. We may be witnessing a return to the tradition of Poincaré and Birkhoff, for Stephen Smale has obtained and continues to obtain important and interesting results in electrical engineering, biology, and economics.

REFERENCES

Birkhoff, George David. "What Is the Ergodic Theorem?" *American Mathematical Monthly* 49 (1942): 222–26.

———. *Collected Mathematical Papers.* American Mathematical Society, 1950.

Gillespie, C. C., ed. *Dictionary of Scientific Biography,* vol. 2. "G. D. Birkhoff." New York: Charles Scribner's Sons, 1970.

Halmos, P. *Ergodic Theory.* New York: Chelsea Publishing Co., 1956.

Morse, M. "George David Birkhoff and His Mathematical Work." *Bulletin of the American Mathematical Society* 52 (1946): 357–91.

Smale, S. "Differentiable Dynamical Systems." *Bulletin of the American Mathematical Society* 73 (1967): 747–817.

———. "What Is Global Analysis?" *American Mathematical Monthly* 76 (1969): 4–9.

25

Fermat's Last Theorem: 1637–1988

CHARLES VANDEN EYNDEN

$\mathcal{A}$ROUND 1637 the French jurist and amateur mathematician Pierre de Fermat wrote in the margin of his copy of Diophantus's *Arithmetic* that he had a "truly marvelous" proof that the equation $x^n + y^n = z^n$ has no solution in positive integers if $n > 2$. Unfortunately the margin was too narrow to contain it. In 1988 the world thought that the Japanese mathematician Yoichi Miyaoka, working at the Max Planck Institute in Bonn, West Germany, might have discovered a proof of this theorem. Such a proof would be of considerable interest because no evidence has been found that Fermat ever wrote one down, and no one has been able to find one in the 350 years since. In fact Miyaoka's announcement turned out to be premature, and a few weeks later articles reported holes in his argument that could not be repaired.

Reports of the latest supposed proof of Fermat's last theorem, as the result has come to be called, seem to have been triggered by a talk given by Miyaoka in Bonn to about three dozen mathematicians on 26 February 1988. Those present started calling friends about his apparent proof, and eventually word reached the newspapers and weekly magazines.

When Miyaoka's proof fell through, members of the mathematical community expressed recriminations, fearing that the public would judge that

Reprinted from *Mathematics Teacher* 82 (Nov., 1989): 637–40; with permission of the National Council for Teachers of Mathematics.

mathematicians could not tell what was a proof and what was not. Some blamed reporters for rushing into print too soon, whereas others blamed mathematicians for talking to reporters before the proof was confirmed. Another view was that any publicity was better than none. At least mathematics was shown to be a living subject, practiced by human beings capable of mistakes like anyone else.

The Early History of Fermat's Equation

The mathematical historian Eric Temple Bell (1961) says that Fermat is the only amateur to reach the first rank in mathematics. Certainly he would have had difficulty achieving recognition in modern times, since he never published a mathematical work of any kind under his name. In fact only one number-theoretic proof by Fermat survives. His famous marginal note was found by his son Samuel, who went through Fermat's letters and unpublished works with an eye toward their posthumous publication.

Probably one reason Fermat's problem has survived is that it is possible to prove it for particular exponents. Thus early workers on the problem might have come up with something publishable, if not a general proof. Fermat himself probably had a proof for $n = 4$, and Gauss succeeded for the more difficult case of $n = 3$. Many famous mathematicians settled other cases. The proof for $n = 5$

was first given by Legendre in 1825, using an idea of Dirichlet. The latter's settling of $n = 14$ in 1832 was superseded by Lamé's 1839 proof that Fermat's equation is impossible for $n = 7$.

Proving Fermat's last theorem for a given exponent n also settles it for any multiple of n. For example, knowing that

$$x^7 + y^7 = z^7$$

is impossible in positive integers also covers the case $n = 14$, since if we had

$$X^{14} + Y^{14} = Z^{14},$$

then

$$x = X^2, y = Y^2, z = Z^2$$

would satisfy the first equation. Thus, to prove Fermat's last theorem in general, it suffices to prove that

$$x^n + y^n = z^n$$

is impossible in positive integers for $n = 4$ and for any value of n that is an odd prime. We can also assume that the greatest common divisor of x, y, and z is 1, since, if d were a common factor, then the equation could be divided by d^n. Such a solution is said to be *primitive*. It is easily seen that for a primitive solution x, y, and z are relatively prime in pairs.

The case $n = 4$ is the only one for which a short proof is known. This proof depends on the integral solutions (called *Pythagorean triples*) to the equation $x^2 + y^2 = z^2$. The general solution to this equation was known by Diophantus and perhaps even by the Babylonians, who before 1600 B.C. at least had a way of generating Pythagorean triples, including 4961, 6480, and 8161 (Neugebauer 1952). The theorem needed is the following. Proofs can be found in Niven and Zuckerman (1980) and Vanden Eynden (1987).

THEOREM. *Suppose the positive integers x, y, and z form a primitive Pythagorean triple. Then one of x and y is even and the other odd. If x is even, then relatively prime positive integers u and v exist such that $x = 2uv$, $y = u^2 - v^2$, and $z = u^2 + v^2$.*

We shall use this theorem to show that Fermat's last theorem holds in the case $n = 4$. Suppose we have a primitive solution to $x^4 + y^4 = z^4$ in positive integers. Setting $w = z^2$ yields the equation

$$x^4 + y^4 = w^2, \tag{1}$$

and so x^2, y^2, w is a primitive Pythagorean triple. Then, assuming x^2 is even, by the theorem just stated, relatively prime positive integers u and v exist such that

$$x^2 = 2uv, y^2 = u^2 - v^2, \text{ and } w = u^2 + v^2.$$

From the second of these equations we have $v^2 + y^2 = u^2$. Since we know that y is odd, we can apply the theorem again to find relatively prime positive integers s and t such that

$$v = 2st, y = s^2 - t^2, \text{ and } u = s^2 + t^2.$$

Note that $x^2 = 2uv = 2(s^2 + t^2)(2st)$, and so

$$(x/2)^2 = st(s^2 + t^2).$$

(Recall that x was even.) Since s and t are relatively prime, the three factors on the right of this equation are relatively prime in pairs. For, if the prime p divides s, for example, then it cannot divide t and so does not divide $s^2 + t^2$ either. It follows from the unique factorization of integers into primes that each of s, t, and $s^2 + t^2$ must itself be a square, say

$$s = x_1^2, \qquad t = y_1^2,$$

and

$$s^2 + t^2 = w_1^2,$$

where x_1, y_1, and w_1 are relatively prime positive integers. But from these equations we have

$$x_1^4 + y_1^4 = w_1^2, \qquad (2)$$

which is of the same form as equation (1). Notice that

$$w_1 \leq w_1^2 = s^2 + t^2 = u \leq u^2 < u^2 + v^2 = w.$$

Thus from (1) we have derived another case of the sum of two fourth powers being a square, with the squared integer smaller than the one we started with.

This example illustrates a type of argument due to Fermat, from the one proof he left, known as the *method of infinite descent*. Starting with equation (2), we could similarly find positive integers x_2, y_2, and w_2 such that

$$x_2^4 + y_2^4 = w_2^4,$$

with $w_2 < w_1$. Continuing in this way we get an infinite sequence of positive integers

$$w > w_1 > w_2 > w_3 > \cdots > 0.$$

Since this result is obviously impossible, so is the original equation

$$x^4 + y^4 = z^4,$$

and Fermat's last theorem has been proved for the case $n = 4$.

The Work of Kummer

The most significant contribution toward a proof of Fermat's last theorem was made by the German mathematician Ernst Eduard Kummer. An often-repeated story says that in the 1840s Kummer thought he had a proof based on doing arithmetic in certain domains of *algebraic integers*. These are complex numbers that behave in many ways like ordinary integers. Gauss already used such numbers in his proof of the case $n = 3$, namely, the complex numbers of the form $a + bp$, where a and b are ordinary integers and p is a root of the equation

$x^2 + x + 1 = 0$. The sum and product of such numbers are of the same form, and concepts of divisibility and primality can be defined for them. The story goes that Kummer's proof assumed that an element of any such domain can always be written as the product of primes in a unique way, a property that in fact holds in some, but not all, cases. Only when Kummer submitted his proof to Dirichlet was his mistake discovered.

Recent research by Harold M. Edwards has thrown doubt on the foregoing story. (Most of the historical material in this article is based on Edwards's excellent book on Fermat's last theorem [1977].) Its source seems to be a talk given in 1910, more than sixty years after the alleged incident, by Hensel, who had gotten it from a third party who was not a mathematician. Kummer's supposed proof has never been found, and in general the story is not supported by contemporary documents.

Records of the Paris Academy from 1847 show that both Lamé and Cauchy believed that factorization in algebraic number domains was unique and that a proof of Fermat's last theorem could be based on this idea. Both realized that unique factorization would have to be proved, however, to make the proof complete. In fact, Kummer had published examples of domains where unique factorization did not hold in an obscure journal three years before, and no written evidence has been found that he had ever assumed that such domains did not exist.

What Kummer did prove was that Fermat's last theorem held for all "regular" primes, which will be defined shortly. The power of Kummer's result is indicated by the fact that the smallest prime that is not regular is 37. Thus the cases $n = 3, 5, 7, 11, 13, 17, 19, 23, 29$, and 31 (and many others) are disposed of all at once. In fact the only primes less than 100 that are not regular are 37, 59, and 67.

The regular primes can be characterized in the following way. First we define the *Bernoulli numbers* $B_0, B_1, B_2 \ldots$ by the equation

$$\left(B_0 + \frac{B_1 x}{1!} + \frac{B_2 x^2}{2!} + \cdots\right)$$
$$\cdot\left(\frac{x}{1!} + \frac{x^2}{2!} + \frac{x^3}{3!} + \cdots\right) = x.$$

By multiplying the expressions on the left as if they were polynomials and equating coefficients of like powers of x, the numbers B_0, B_1, ... can be computed recursively. The reader may want to check that $B_0 = 1$, $B_1 = -1/2$, $B_2 = 1/6$, $B_3 = 0$, and $B_4 = -1/30$. It turns out that B_i is always a rational number, and $B_i = 0$ for values of i that are odd and greater than 1. We say that a prime p is *regular* when it does not divide the numerator of any of the Bernoulli numbers B_2, B_4, ... , B_{p-3}. For example, 7 is regular because it does not divide the numerators of B_2 or B_4.

Unfortunately neither Kummer nor anyone since has been able to prove the existence of infinitely many regular primes. The existence of infinitely many primes that are not regular has been proved.

Results after Kummer

Although many theorems have been proved since Kummer on the subject of Fermat's last theorem, most cannot be stated simply. Many amateurs were encouraged to work on the problem by a prize of 100 000 marks offered for a proof (but not a disproof) of the theorem in the will of Paul Wolfskehl in 1909. The proof had to be accepted as valid by the German academy of sciences in Göttingen. Inflation after World War I reduced the value of the prize to almost nothing, but it is now worth about $5500. The Mathematics Institute of the University of Göttingen still receives three or four "proofs" per month aiming at the prize.

In modern times computers have been turned loose on the problem. Samuel Wagstaff and Jonathan Tanner (1987) have used computer techniques to prove the theorem for all exponents $n \le$ 150 000. A recent theoretical advance was the proof of Gerd Faltings that for any fixed value of n, Fermat's equation can have only a finite number of primitive solutions.

Methods of proof are now applied that go far beyond anything Fermat could have imagined. These new approaches make it increasingly difficult to believe that Fermat really had a proof.

BIBLIOGRAPHY

Bell, Eric Temple. *The Last Problem*. New York: Simon & Schuster, 1961.

Dembart, Lee. "Scientists Buzzing—Fermat's Last Theorem May Have Been Proved." *Los Angeles Times*, 8 March 1988, 3, 23.

*Dickson, Leonard Eugene. *History of the Theory of Numbers*. Vol. 2. New York: Chelsea Publishing Co., 1952.

Edwards, Harold M. *Fermat's Last Theorem*. New York: Springer-Verlag, 1977.

Hilts, Philip J. "Famed Math Puzzle May Have Been Solved." *Washington Post*, 9 March 1988, 1, 4.

Kolata, Gina. "Progress on Fermat's Famous Math Problem." *Science* 235 (27 March 1987): 1572–73.

Mordell, Louis J. *Three Lectures on Fermat's Last Theorem*. Cambridge: Cambridge University Press, 1921.

*Neugebauer, Otto. *The Exact Sciences in Antiquity*. Princeton University Press, 1952.

Niven, Ivan, and Herbert S. Zuckerman. *An Introduction to the Theory of Numbers*. 4th ed. New York: John Wiley & Sons, 1980.

Ribenboim, Paulo. *13 Lectures on Fermat's Last Theorem*. New York: Springer-Verlag, 1979.

Tanner, Jonathan W., and Samuel S. Wagstaff, Jr. "New Congruences for the Bernoulli Numbers." *Mathematics of Computation* 48 (1987): 341–50.

Vanden Eynden, Charles. *Elementary Number Theory*. New York: Random House, 1987.

*Available as a Dover reprint.

EDITOR'S NOTE:

Fermat's Last Theorem, 1993

On June 23, 1993, Dr. Andrew Wiles, of Princeton University, delivered the third of a series of conference lectures at the Newton Institute, Cambridge University. Dr. Wiles's lecture series was entitled "Modular Forms, Elliptic Curves, and Galois Representations" and described his research findings associating the theory of elliptic curves, that is, curves designated by equations of the form $y^2 = x^3 + ax + b$, where a and b are integers, with Fermat's Last Theorem. Wiles proved a conjecture which originated in 1955 with the Japanese mathematician Yutaka Taniyama and was later modified by Andre Weil (1968) and Goro Shimura (1971). It is known as the Shimura-Taniyama-Weil conjecture and associates the behavior of elliptic curves with another class of curves known as modular curves. In 1986, Kenneth Ribet of the University of California found a link between elliptic curves and Fermat's Last Theorem. He showed that if a counterexample existed to Fermat's Last Theorem then it would be possible to construct an elliptic curve that was not modular. Building upon Ribet's theories and employing methods developed by still other researchers, Wiles established the Shimura-Taniyama-Weil conjecture for a class of elliptic curves including some relevant to proving Fermat's Last Theorem. In accomplishing this feat, he, in principle, proved Fermat's Last Theorem and established solution techniques that will greatly advance number theory research.

In commenting upon Andrew Wiles's achievement, Barry Mazur, a number theorist at Harvard University who himself had been involved in research on the subject, noted that although the problem was an ancient one solved with modern techniques, still it

> was answered strictly in its own terms. That's rather amazing right there. It underscores how stable mathematics is through the centuries—how mathematics is one of humanity's long continuous conversations with itself.

Wiles's victory certainly testifies to his personal genius and dogged persistence—he worked on the problem for seven years. He deserves the laurels of the victor but, in a larger sense, his accomplishment is also a victory for the mathematical community and demonstrates international cooperation and interdependence in the field of mathematical research. In developing his solution scheme, Andrew Wiles employed theories from many branches of mathematics: crystalline cohomology, Galois representations, L-functions, modular forms, deformation theory, Gorenstein rings . . . and relied on research findings from colleagues in France, Germany, Italy, Japan, Australia, Colombia, Brazil, Russia, and the United States. Wiles, himself, is a British citizen working at an American university. This relay race has been won.

—FRANK SWETZ

The Tree of Mathematics

Dr. Margie Hale, Professor of Mathematics at Stetson University, Florida, shares her enthusiasm for mathematics through the following analogy:

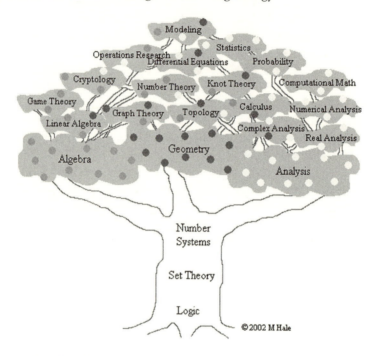

I conceive of mathematics as a fantastic citrus tree: the three main branches of oranges, limes, and lemons representing the major fields of algebra, geometry, and analysis. Each part of the upper canopy takes advantage of what is below it. Thus, number theory has two great approaches, one algebraic and the other analytic. Topology makes use of all three fields, as does the very different field of mathematical modeling.

All trees have woody parts, foliage, and fruit. The trunk and branches absorb water and nutrients from the soil. The foliage absorbs energy from the sun, using it to convert the nutrients to usable food, and allowing the tree to grow. The fruit allows the tree to reproduce. I have represented the foundations of mathematics as the trunk of the tree, the support and food supply. The foliage and fruit could represent the two aspects of pure and applied mathematics: which is which, do you think?

Epilogue:
Mathematics, A Living Organism

*A*N EPILOGUE is usually an afterthought added to a written work to reflect on its contents or, at a later instance, to modify some of its conclusions. In a sense, it stands apart from the textual message itself. In the case of this epilogue, it is very much a part of the textual message of just how mathematics works.

When I completed the compilation of From Five Fingers to Infinity in 1991, I purposefully ended with Vanden Eynden's article on Fermat's Last Theorem, the most famous unsolved problem in the history of mathematics. His article highlights some of the uncertainty in mathematics: an unsolved problem, an almost correct solution, a continued search for a solution, and illustrates the persistent nature of mathematicians in striving to solve the problem of Fermat's Last Theorem. In this mission, the work of one mathematician supplies a basis for a latter colleague to carry on the quest. Teamwork is evident. The research process of seeking a solution to a difficult mathematical problem is like a relay-race with each participant carrying the baton a little closer to the victory line before handing it off to his or her teammate. Then as the manuscript was being prepared for publication, in June of 1993, Andrew Wiles, of Princeton University announced his solution of Fermat's Last Theorem. How convenient! I then continued to discuss the background of Wiles' work and the interdisciplinary and cooperative endeavor of the solution process within the mathematics community. This example demonstrated the persistent nature of mathematical inquiry and growth; and, I thought, provided a fitting final statement for the book.

In the process of this recent revision, the question of an epilogue again rose up. Was it necessary? Yes, because I feel the history of mathematics is never complete and perhaps the reader should be left with a direction to seek more knowledge about this subject. Following the precedent of the previous edition, I sought another unsolved mathematical problem to pursue. I then came across Margie Hale's tree analogy and my dilemma was solved.

Mathematics is a living, thriving structure. Rooted in reality, the "natural world," like a tree it rises up nurtured by human need, curiosity and intellectual genius. New areas of research (branches) are constantly produced and flourish. Most recently subjects such as fractal geometry, fuzzy logic, chaos theory and string theory have blossomed and offer potential for better understanding and control of the environment within which we exist. New insights pertaining to applications are being obtained in number theory and classical analysis. The journey through mathematics the reader has just taken has briefly traced out the germination and initial growth of mathematics over a relatively brief period of history. That growth is continuing. One can only wonder where it will lead us.

Suggested Readings

For a broad perspective on mathematics from 1800 to 2008:

Katz, Victor J. *A History of Mathematics: Brief Edition.* New York: Pearson/Addison Wesley, 2003. Chapters 16–20.

Suzuki, Jeff. *A History of Mathematics.* Upper Saddle River, NJ: Prentice Hall, 2002. Chapters 19–24.

For more specialized examinations:

Barker, Stephen F. *Philosophy of Mathematics.* New York: Prentice Hall, 1964.

Black, Max. *The Nature of Mathematics; A Critical Survey.* New York: Humanities Press, 1950.

Crowe, M. J. *A History of Vector Analysis.* (1967 reprint). New York: Dover Publications, 1994.

Dauben, Joseph Warren. *Georg Cantor: His Mathematics and Philosophy of the Infinite.* Princeton, NJ: Princeton University Press, 1979.

Davis, Philip. *The Mathematical Experience.* Boston: Birkhäuser, 1981.

Gratton-Guinness, I. *The Development of the Foundations of Mathematical Analysis from Euler to Riemann.* Cambridge, MA: MIT Press, 1970.

Greenberg, Marven. *Euclidean and Non-Euclidean Geometry: Development and History.* San Francisco: W. H. Freeman, 2008.

Hairer, E. and Wanner G. *Analysis by Its History.* New York: Springer-Verlag, 1996.

Ifrah, Georges. *The Universal History of Computing: From the Abacus to the Quantum Computer.* New York: John Wiley & Sons, 2001.

Johnson, P. E. *A History of Set Theory.* Boston: Prindle, Weber & Schmidt, 1972.

Kleiner, Israel. *A History of Abstract Algebra.* Boston: Birkhäuser, 2007.

Kline, Morris. *Mathematics: The Loss of Certainty.* New York: Oxford University Press, 1979.

Meschkowski, Herbert. *Ways of Thought of Great Mathematicians.* Translated by John Dyer-Bennet. San Francisco: Holden-Day, 1964.

Odifreddi, Piergiorgio. *The Mathematical Century: The 30 Greatest Problems of the Last 100 Years.* Princeton, NJ: Princeton University Press, 2004.

Singh, Simon. *Fermat's Enigma: The Epic Quest to Solve the World's Greatest Mathematical Problem.* New York: Anchor Books, 1997.

Styazhkin, N. I. *History of Logic from Leibniz to Peano.* Cambridge, MA: MIT Press, 1969.

Yandell, Ben H. *The Honors Class: Hilbert's Problems and Their Solvers.* Natick, MA: A. K. Peters, Ltd., 2002.

Index